# Hardware Supply Chain Security

Basel Halak
Editor

# Hardware Supply Chain Security

## Threat Modelling, Emerging Attacks and Countermeasures

*Editor*
Basel Halak
Electronics and Computer Science School
University of Southampton
Southampton, UK

ISBN 978-3-030-62709-6 ISBN 978-3-030-62707-2 (eBook)
https://doi.org/10.1007/978-3-030-62707-2



This Springer imprint is published by the registered company Springer Nature Switzerland AG
The registered company address is: Gewerbestrasse 11, 6330 Cham, Switzerland

Basel Halak
Editor

# Hardware Supply Chain Security

## Threat Modelling, Emerging Attacks and Countermeasures

*Editor*
Basel Halak
Electronics and Computer Science School
University of Southampton
Southampton, UK

ISBN 978-3-030-62709-6    ISBN 978-3-030-62707-2 (eBook)
https://doi.org/10.1007/978-3-030-62707-2



This Springer imprint is published by the registered company Springer Nature Switzerland AG
The registered company address is: Gewerbestrasse 11, 6330 Cham, Switzerland

*To*
*Suzanne, Hanin and Sophia*
*with Love*

# Preface

The trend towards globalisation and the need to cut costs to gain competitive advantages have resulted in a remarkable growth of outsourcing levels; this is particularly true for the hardware supply chain. The latter has become a multinational distributed business. This evolution of the supply chain structure has brought about a number of serious challenges, including a rising level of IP piracy, counterfeiting and the emergence of new forms of attacks such as Hardware Trojans. Such attacks can lead severe consequences. Financially, counterfeiting is costing the global economy billions of US dollars every year. Furthermore, compromised hardware products pose a serious security threat if used in critical infrastructure and military applications. The threat on the security of hardware devices is exasperated by the proliferation of the internet of things (IoT) technology, wherein the majority of devices have limited computation and memory resources, making it harder to implement typical security defence mechanisms. In fact, in 2019, a staggering 2.9 billion cyberattacks on internet of things devices have been recorded. What is more, a significant portion of such embedded systems are deployed in unprotected and physically accessible locations, therefore they can be vulnerable to both invasive and side channel attacks, allowing attackers to extract sensitive data that might be stored on IoT nodes, such as encryption keys, digital identifiers and recorded measurements. To mitigate against such risks, engineers need to treat security as an integral part of the design process, not as an after-thought.

Overlooking security in the electronic products development will put a great many systems at risk, leading, in many cases, to financial losses, damage of reputation and in extreme cases to physical harm.

Building effective defence mechanisms requires a comprehensive understanding of the attack classes, goals of the adversary and their capabilities. However, what makes devising appropriate countermeasures particularly challenging is that fact that an electronic system does not always recognise it is under attack, so it may fail to activate its defences at the right time. This is because, in many cases, the attack is new, so its symptoms are unknown or the system attributes the attack to a reliability problem. Therefore, the detection of anomalous behaviours in electronic systems is an important defence technique.

The prime objective of this book is to provide a timely and a comprehensive account of emerging security attacks on the hardware supply chain, which covers all stages of an electronic system's life cycle, from initial design specifications, through its operation in the field, until it is discarded in an IC recycling centre.

To facilitate the understanding of the material, each chapter includes background information explaining related terminologies and principles, in addition to a comprehensive list of relevant references. The book is divided into three parts to enhance its readability, namely threat modelling of Hardware supply chain, Emerging attacks and countermeasures and anomaly detection in embedded systems.

## The Contents at Glance

This book presents a new threat modelling approach that specifically targets hardware supply chain, covering security risks throughout the lifecycle of an electronic system. Afterword's the book presents a case study on a new security attack, which combines two forms of attack's mechanisms, from two different stages of the IC supply chain, more specifically the attack targets the newly developed light cypher (Ascon) and demonstrates how it can be broken easily, when its implementation is compromised with a hardware Trojan.

The book also discusses emerging countermeasures, including anti-counterfeit design techniques for resources-constrained devices and anomaly detection methods for embedded systems. More details on each chapter are provided below.

Part I: Threat Modelling of Hardware Supply Chain

Chapter 1 systemises the current knowledge on hardware security including emerging attacks and state-of-the-art defences, and presents a number of case study to demonstrate how to perform security validation of countermeasures in the context of specific application scenario, which allows balancing security requirements of a particular product with the cost associated with implementing defence mechanisms.

Part II: Emerging Hardware-Based Security Attacks and Countermeasures

Chapter 2 provides a case study on an emerging security threat that combines multiple attack's vectors. It demonstrates the feasibility of using hardware Trojan to undermine the resilience of Ascon cyphers to the cube attacks. The combined attack was successfully waged on a FPGA implementation of the Ascon cypher.

Chapter 3 focuses on the security of resources-constrained systems, in particular RFID. It presents a new defence mechanism to mitigate against counterfeiting attacks, such as tag cloning. This chapter proposes a new security mechanism, which consists of a lightweight three-flight mutual authentication protocol and an anti-counterfeit tag design. The solution is based on combining the Rabin public key encryption scheme with physically unclonable functions (PUF) technology.

Part III: Anomaly Detection in Embedded Systems

Chapter 4 outlines the underlying causes of anomalous behaviours in embedded systems, distinguishing between those caused by reliability problems and others resulting from security attacks.

Chapter 5 provides an overview of hardware performance counters, an important tool used for anomaly detection in electronic systems.

Chapter 6 provides a comprehensive summary of anomaly detection techniques and presents a detailed case study on the use of machine learning algorithms in this context.

## Book Audience

The book is intended to provide a comprehensive coverage of the latest research advances in the key research areas of hardware supply chain security; this makes it a valuable resource for graduate student researchers and engineers working in these areas. I hope this book will complement ongoing research and teaching activities in this field.

Southampton, UK
September 2020

Basel Halak

# Acknowledgments

I would like to thank all of those who contributed to the emergence, creation and correction of this book. Firstly, I gratefully acknowledge the valuable contributions from my students at the University of Southampton, for the many hours they have spent working in their labs to generate the experimental results. Of course, the book would not be successful without the contributions of many researchers and experts in the field of security and embedded systems.

Finally, I would like to thank the great team at Springer for their help and support throughout the publication process.

# Contents

# About the Editor

**Basel Halak** is the director of the embedded systems and IoT programme at the University of Southampton, a visiting scholar at the Technical University of Kaiserslautern, a visiting professor at the Kazakh-British Technical University, an industrial fellow of the royal academy of engineering and a senior fellow of the higher education academy. He has written over 70-refereed conference and journal papers, and authored four books, including the first textbook on Physically Unclonable Functions. His research expertise includes evaluation of security of hardware devices, development of appropriate countermeasures, the development of mathematical formalisms of reliability issues in CMOS circuits (e.g. crosstalk, radiation, ageing), and the use of fault tolerance techniques to improve the robustness of electronics systems against such issues. Dr. Halak lectures on digital design, Secure Hardware and Cryptography, supervises a number of MSc and PhD students, and is the ECS Exchange Coordinators. He is also leading the European Masters in Embedded Computing Systems (EMECS), a 2-year course run in collaboration with Kaiserslautern University in Germany and the Norwegian University of Science and Technology in Trondheim (electronics and communication). Dr. Halak serves on several technical programme committees such as HOST, IEEE DATE, IVSW, ICCCA, ICCCS, MTV and EWME. He is an associate editor of IEEE access and an editor of the IET circuit devices and system journal. He is also a member of the hardware security-working group of the World Wide Web Consortium (W3C).

# Part I
# Threat Modelling of Hardware Supply Chain

# Chapter 1
# CIST: A Threat Modelling Approach for Hardware Supply Chain Security

**Basel Halak**

## 1.1 Introduction

### 1.1.1 Motivation

The remarkable growth of outsourcing in the hardware supply chain has brought about serious challenges in the form of new security attacks, particularly IC counterfeit and Hardware Trojan insertion. Such attacks can have severe consequences. Financially, counterfeiting is costing the UK economy around £30 billion and is putting 14,800 jobs at risk [1]. The consequences of an insecure IC supply chain are not only limited to major financial losses, but they also pose a national security threat. A recent report by Bloomberg alleged that Chinese spies reached almost 30 US companies, including Amazon and Apple, by compromising America's technology supply chain by inserting a hardware Trojan.[1] Although the above-mentioned companies have denied these claims, an analysis by a prominent hardware security expert indicated the above attack was entirely feasible.[2] Another study has indicated that counterfeit ICs can cause the malfunctioning of military weapons and vehicles [2]. Another trend that has further increased the hardware attack surface is the proliferation of the internet of things (IoT) technology. The

---

[1] https://www.bloomberg.com/news/features/2018-10-04/the-big-hack-how-china-used-a-tiny-chip-to-infiltrate-america-s-top-companies.

[2] https://www.google.co.uk/amp/s/gigazine.net/amp/en/20181010-supermicro-motherboard-attack.

---

B. Halak (✉)
Electronics and Computer Science School, University of Southampton, Southampton, UK
e-mail: basel.halak@soton.ac.uk


B. Halak (ed.), *Hardware Supply Chain Security*,
https://doi.org/10.1007/978-3-030-62707-2_1

latter has led to a rapid increase in computing devices (around 50 billion in 2025). In fact, in 2019, a staggering 2.9 billion cyberattacks on IoT devices have been recorded [3].

This steady drumbeat of news stories on successful security attacks makes it abundantly clear that current defence mechanisms are insufficient to protect from the developing risks of cyber and physical attacks. One reason behind the failure of current practices is that the threat models on which the security of systems rely fail to account for the big picture [4].

Threat modelling is a process by which potential threats, such as structural vulnerabilities or the absence of appropriate safeguards, can be identified, enumerated and mitigations can be prioritized. Such a process makes it feasible to perform systematic analysis of the probable attacker's profile; the most likely attack vectors, the assets most desired by an attacker, and more importantly the most effective defence mechanisms.

Although, there are plenty of work in the literature on the different types of attacks on hardware supply chain and computing devices, there are very few attempts to develop a comprehensive threat modelling approach that considers the life cycle of hardware systems from the initial design specifications to the recycling phase of discarded ICs.

One of the earliest work in this area is [5], in which the need for systematic approach to analysing the security threat of the IC supply chain is highlighted. A more comprehensive security analysis appeared few years later in [6], in which the authors provided a summary of exiting attacks and possible countermeasures. Both of these papers indicated that existing threat modelling approach such as those based on the Microsoft's STRIDE model [7, 8] is not suitable for analysing the security of IC supply chain; this is because the type of threats in the latter case is of an inherently different nature compared to those included in the STRIDE model. To the best of our knowledge, there is currently no comprehensive threat modelling approach that is tailored to the needs of IC supply chain. Furthermore, new IC fabrication techniques are continuously being developed to keep pace with Moore's law [9]; these new approaches may have fundamentally different security assumptions, which need to be systematically evaluated and analysed in order to understand whether or not these new models carry any unforeseen security risks. Having a clear understanding of the types of threats of future IC supply chain will allow for defence mechanisms to be put in place at an early stage of the development process, i.e. building a secure-by-design IC supply chain.

The contributions of this work are as follows:

1. It systematizes the current knowledge on security attacks on hardware supply chain, providing a comprehensive review of security attacks and the state-of-the-art countermeasures.
2. It presents a new threat modelling approach specifically developed for hardware supply chain, which takes into consideration the nature of the system, the most probable attacker's profile and the root of vulnerability. This allows the development of appropriate countermeasures that balances security needs with available resources and other operation requirements.

3. It presents a number of exemplar case studies that demonstrate how the proposed threat modelling approach can be used to evaluate the appropriateness and the level of protection provided existing defence mechanisms in the context of a specific application.

### *1.1.2 Chapter Summary*

The remainder of this paper is organized as follows. Section 1.2 reviews related work and preliminary background. Sections 1.3 and 1.4 present a summary of the proposed threat modelling approach called *CIST*. Section 1.5 explains the (PROV-N) technique adopted in this work for modelling the IC supply chain security. Sections 1.6 and 1.7 provide a comprehensive review of the attacks on the hardware supply chain and corresponding countermeasures, respectively. Section 1.7 demonstrates using a number of case studies how the CIST approach can be used for security validation of existing countermeasures. Conclusions are drawn in Sect. 1.9.

## 1.2 Background

### *1.2.1 Related work*

Modelling of security threats and attacks has been extensively studied in the literature. Attack trees proposed in [10] provide a systematic way of describing the security of a system, each tree consists of one root, leaves and children. The root is the ultimate goal of the attack (e.g. insert a Trojan). The child nodes are conditions that must be satisfied to make the direct parent node true. In other words, the attack described in the root may require one or more of many attacks described in child nodes to be satisfied. Attack pattern [10] is a similar approach that relies on the use of AND/OR composition of operations (Pattern) that may generate an attack. Both methods are originally developed to analyse software and networking system security, therefore, their use to model hardware security threat has not been investigated.

The work in [11] proposed a unified conceptual framework for security auditing from a risk management perspective through the generation of threat models. This approach, called "Trike", relies on a "requirements model." The latter is used to ensure the level of risk assigned to each asset is appropriate to relevant stakeholders. This can be challenging to implement in large complex application scenarios, as it needs the designer to have an overview of the entire system to be able to conduct the attack surface analysis. Another widely used approach is STRIDE [7], which is used to categorize the identified threats into six categories *Spoofing, Tampering, Repudiation, Information disclosure, Denial of service* and *Elevation of privilege*. The goal of this approach is to ensure that applications meet the security properties of *Confidentiality, Integrity and Availability (CIA), along with Authorization, Authentication and Non-Repudiation*. All of the above methods are very useful in performing threat and risk analysis. However, these have not been specifically developed for Hardware attacks. Some categories are not applicable to hardware threats, e.g. Spoofing, Repudiation and Elevation of privilege; some can be applied but may have different meanings, e.g. confidentiality and tampering, while some hardware-related risks are not even covered such as IC counterfeit and supply chain sabotage, hence the need for a more tailored threat modelling approach that consider all hardware-related attacks throughout the cycle of the IC.

### *1.2.2 Description of IC Production Process*

Semiconductor technologies have infiltrated all areas of modern life driven by a multitude of emerging applications. The unprecedented demand for cheaper and more complex silicon chip has led to a rise in the outsourcing level in IC production process. The latter has become a multinational distributed business that involves hundreds of suppliers and complex logistics. Figure 1.1 provides a summary of the IC production chain. The first stage consists of sourcing (IP) designs from third party providers. The second stage is the system-on-chip (SoC) integration, which takes place at the design house and produces the layout files. The latter are sent for fabrication and testing. The next stage is packaging and integration into the final product. The latter will be in use for a period of time, depending on its nature (e.g. mobile phone are used for 2–3 years, TV are used for 5–10 years . . . ). Once the product is discarded, some silicon chips are illegally recycled to be used in other products.

### *1.2.3 Preliminaries*

#### 1.2.3.1 How to Develop a Secure System in a Nutshell

The notion of "security" is relative. If an adversary has unlimited amount of time and resources, they can break any system. "Secure" simply means that the amount of efforts and costs for breaking a system exceeds potential benefits.

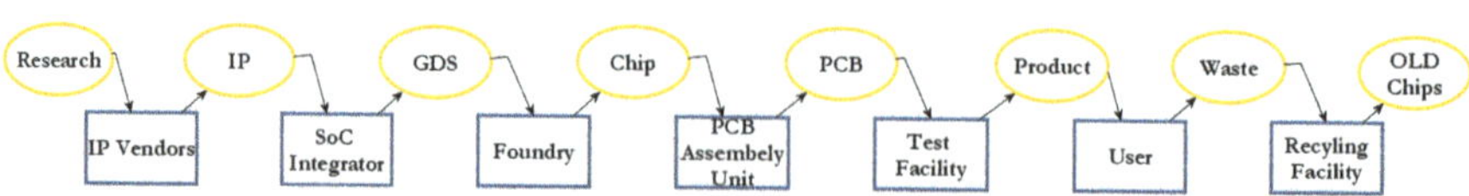


**Fig. 1.1** A simplified illustration of the IC production chain

Therefore, establishing whether or not a product is secure requires understanding the underlying motivation of the most likely adversary. For example, if an attacker was seeking an economic gain, then an effective countermeasure should make breaking the system more expensive than any potential financial benefits. On the other hand, if the adversary's goal were to use the attack as weapon to undermine enemies' capabilities and infrastructure, then they would likely have access to a great deal of recourses, which makes implementing a countermeasure much more difficult, and in some cases futile. The best approach in this case is to prevent the attack by isolating and physically protecting the system or ideally by removing the root of vulnerability that made the attack technically feasible.

The notion of security is "temporary". In other words, the attack surface is constantly changing with new attacks emerging every day. Therefore, a product that was deemed secure at the design time can still be compromised if new attacks mechanisms are developed. Therefore, security defence mechanisms should be continually reviewed and updated if necessarily. However, enhancing defence measures to protect against a newly discovered hardware vulnerability may not be feasible unless this can be done at the software level.

#### 1.2.3.2 Attack Difficulty

Attack difficulty refers to the amount of resources and the level of expertise required to carry an attack successfully. This metric can be particularly useful when assessing the level of protection afforded by various security defence mechanisms (i.e. how much more difficult the attack has become after implementing the countermeasure).

The attack difficulty in this work will be based on the classification proposed in [12] as shown in Table 1.1.

#### 1.2.3.3 Adversary Classification

Adversary can be classified according to their knowledge, resources and motivation into four categories based on [13] as shown in Table 1.2. Knowing the class of the most likely adversary is vital for developing effective countermeasures.

## 1.3 CIST: A Hardware-Specific Threat Modelling Approach

The proposed approach covers hardware-related risks throughout the life cycle of the IC from design to recycle.

The proposed modelling process is comprised of five high-level steps similar to that in as shown in Fig. 1.2.

The aim of the first step is to define the desired hardware security properties, which can be summarized as *authenticity, confidentiality, integrity* and *availability*.

**Table 1.1** Classification of attack difficulty

| Level | Name | Description |
|---|---|---|
| 1 | Common tools | Commonly available tools and skills may be used (e.g. those tools available from retail department or computer stores, such as a soldering iron or security driver bit set). |
| 2 | Unusual tools | Uncommon tools and skills may be used, but they must be available to a substantial population (e.g. multi-meter, oscilloscope, logic analyser, hardware debugging skills, electronic design and construction skills). Typical engineers will have access to these tools and skills. |
| 3 | Special tools | Highly specialized tools and expertise may be used, as might be found in the Laboratories of universities, private companies or governmental facilities. The attack requires a significant expenditure of time and effort. |
| 4 | In laboratory | A successful attack would require a major expenditure of time and effort on the part of a number of highly qualified experts, and the resources available only in a few facilities in the world. |
| 5 | Not feasible | The attack is no longer feasible |

**Table 1.2** Classes of the adversaries

| Resources | Class 1 (a small group of curious hackers) | Class 2 (an academic research group) | Class 3 (an organized criminal gang) | Class 4 (a state-funded organization) |
|---|---|---|---|---|
| Time | Limited | Moderate | Large | Large |
| Budget | <$10,000 | <$500,000 | >$500,000 | Unknown but Large |
| Goal | Challenge/prestige | Publicity | Money | Varies |
| Organized? | No | Yes | Yes | Yes |
| Will they release the attack's information? | Yes | Yes | No | No |
| Attacks they can wage | 1–2 | 1–3 | 1–3 | 1–4 |

The "model" step will develop a graphical representation of the life cycle of an integrated circuit, which describes its main processes, their interactions and related security risks.

The "Identify Threats" step aims to provide an answer to the question "What do Attackers want". Understanding the goal of the attackers is critical to developing appropriate countermeasures. To achieve this, the following clarification of the goals of the IC supply chain attacker is proposed and adopted in this work, these are tied closely with security properties discussed previously, namely *counterfeiting, information Leakage, sabotage and tampering* (CIST). This step will also identify the attacks mechanisms related to each threat category. This includes identifying the following information for each attack:

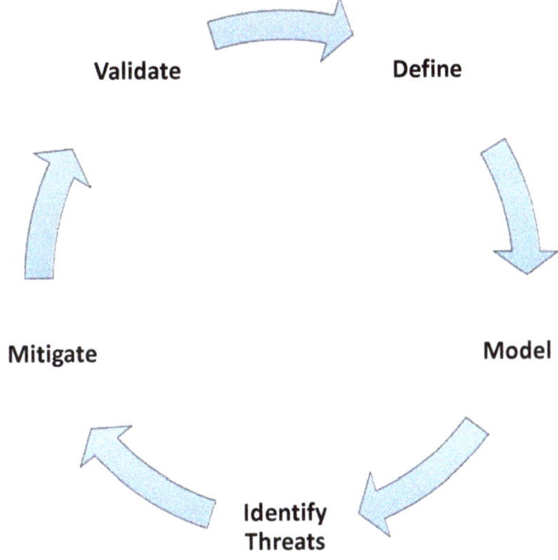


**Fig. 1.2** The Threat Modelling Process of IC Supply Chain

1. The most likely adversary to wage the attack.
2. The attack difficulty.
3. The preconditions required to perform it. This include the assumed capabilities of the adversary and the root of security vulnerability. The latter include the factors that makes the attack feasible, technically or otherwise.
4. The stage(s) at which it is likely to take place during the IC life cycle.
5. Its impact or the damage it can cause (e.g. financial losses and reputation damage political gains).

The "mitigate" step identifies the defence techniques that could be used to mitigate the risk of ach attack.

The "validate" step considers the effectiveness of each of the countermeasures identified in step 4, taking into consideration the profile of the most likely attacker, the difficulty of the attack after implementing the countermeasure and the application scenario.

The CIST approach as identified above allows for systematic analysis of security threats and the identification of most appropriate countermeasures, given a particular application scenario.

## 1.4 Define: Definition of Security Properties

There are four fundamental goals for hardware security, which are based on the IC's applications and life cycle, namely:

- Authenticity is the assurance that an integrated circuit, an embedded device or any other piece or hardware is from the source it claims to be from. Authenticity

**Table 1.3** CIST: a threat modelling approach for the IC supply chain

| Threat | Security property | Definition | Examples of attack mechanisms |
|---|---|---|---|
| Counterfeiting | Authenticity | Fraudulently imitating an original IC | IC recycling<br>IC overproduction |
| Information leakage | Confidentiality | Exposing sensitive design information or secret data stored on chip | Side-channel analysis<br>IP piracy |
| Sabotage | Availability | Deliberately damage or destroy an IC or obstruct its production | Cyber-physical attacks<br>Attacks on transport links to delay IC production |
| Tampering | Integrity | Maliciously change the data associated with the IC, during its development process or after its deployment | Trojan insertion<br>Fault injections |

requires a proof of identity, which should be provided by the device being checked.

- Confidentiality is the property that information is not made available or disclosed to unauthorized individuals, entities or processes. In the context of hardware security, this information includes any sensitive data or code stored in the device or design secrets related to the intellectual property (IP) of the underlying circuitry.
- Integrity: means maintaining and assuring the accuracy, correct functionality and trustworthiness of the hardware device over its entire lifecycle, in other words the device is doing what it is expected to do nothing less and nothing more.
- Availability means the integrated circuit and its different abstraction levels are available when it is needed at every stage of its life cycle, from the initial specification to the final deployed hardware device. An example of an attack that undermine the availability include sabotaging the manufacturing cite to delay production or tampering with a deployed system to take it out of service. All types of Hardware security attacks aim to undermine one or more of the above four properties.

Table 1.3 provide a summary of the threats and corresponding security properties and exemplar attacks.

## 1.5 Model: Modelling the IC Supply Chain Security

There are a number of approaches that can be adopted here, data flow diagram [7] and PROV-N [14]. PROV-N is the notation of W3C provenance PROV family, which brings information about activities and people involved in producing a piece of data. This information can be used to assess quality, reliability and trustworthiness [15]. PROVDM is the conceptual data model and distinguishes core structures, forming the essence of provenance information, from extended structures catering for more specific uses of provenance. PROV-DM is organized into six components, respectively, dealing with: (1) entities and activities, and the time at which they were created, used, or ended; (2) derivations of entities from entities; (3) agents bearing responsibility for entities that were generated and activities that happened; (4) a notion of bundle, a mechanism to support provenance of provenance; and (5) properties to link entities that refer to the same thing; (6) collections forming a logical structure for its members. This semantic modelling approach lends itself well to the description of the IC supply chain in the context of threat modelling, as it makes it feasible to construct a detailed articulation of the stages of the IC production, the agents responsible on each stages and associated security threats. Furthermore, it makes it feasible to investigate the vulnerabilities that are not yet exploited and expose potential attack combinations, with the lack of empirical data on some of the hardware attacks, the hypothetical attack scenarios provided by semantic modelling can assist in developing appropriate security policies for the IC supply chain. Therefore, the semantic modelling used in this research is based on the PROV-N modelling [16]. This was adapted to first model the IC manufacturing production line. The workflow was analysed from general description of IC manufacturing and was mapped using the following rules.

*Rule 1 a process uses an entity to generate another entity.*
*Rule 2 an entity is either an input or an output.*
*Rule 3 a stakeholder is a person(s) that is directly responsible for a process.*

Additional rules for the threat analysis are generated by blending the cyber-physical taxonomy [17] with semantic modelling using Prov-N as follows:

*Rule 4 an attack targets only an entity that was generated by a process.*
*Rule 5 an attack that invalidates a generated entity automatically invalidates the adjacent process that used the same entity.*
*Rule 6 an attack needs a precondition for the attack to be successful.*
*Rule 7 an attack is carried by an attacker.*

The graphical representation of the model will use the coloured notations listed in Table 1.4:

The above rules are used to model to each stage of the IC life cycle as shown in Fig. 1.3.

To explain, how this modelling approach can be used to systematically analyse hardware security threat, we consider a Trojan insertion attack on an IP development company as shown in Fig. 1.4. In this example, the stakeholder is the IP vendor. The latter is responsible for the IP development (i.e. a process). The input for this process is the research performed by the company and associated expenses. The output is the IP. The precondition that need to be satisfied for this attack to succeed is to have access the IP being developed. Such access can be achieved in a number of ways, typically referred to as the *Entry Points* (i.e. the entrances

**Table 1.4** Notation for the CIST threat modelling analysis

| Object | Notation |
|---|---|
| Process | □ |
| Stakeholder | ◇ |
| Entity | o |
| Attack | □ |
| Attacker | △ |
| Preconditions | o |
| Impact | o |

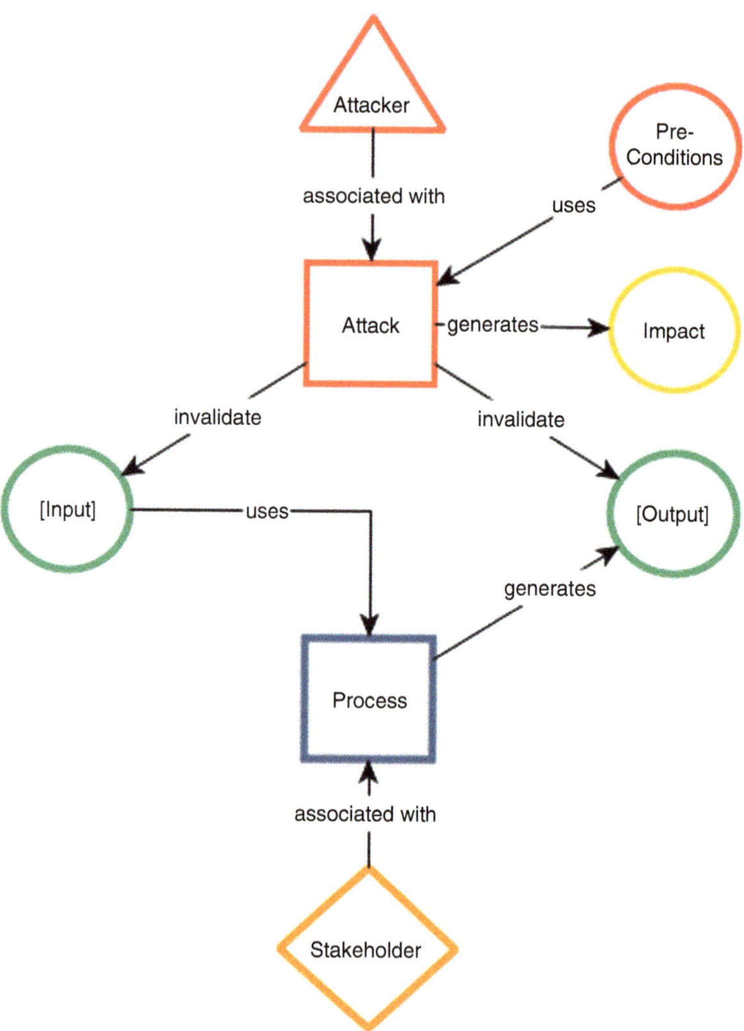


**Fig. 1.3** A reference diagram for hardware supply chain threat modelling

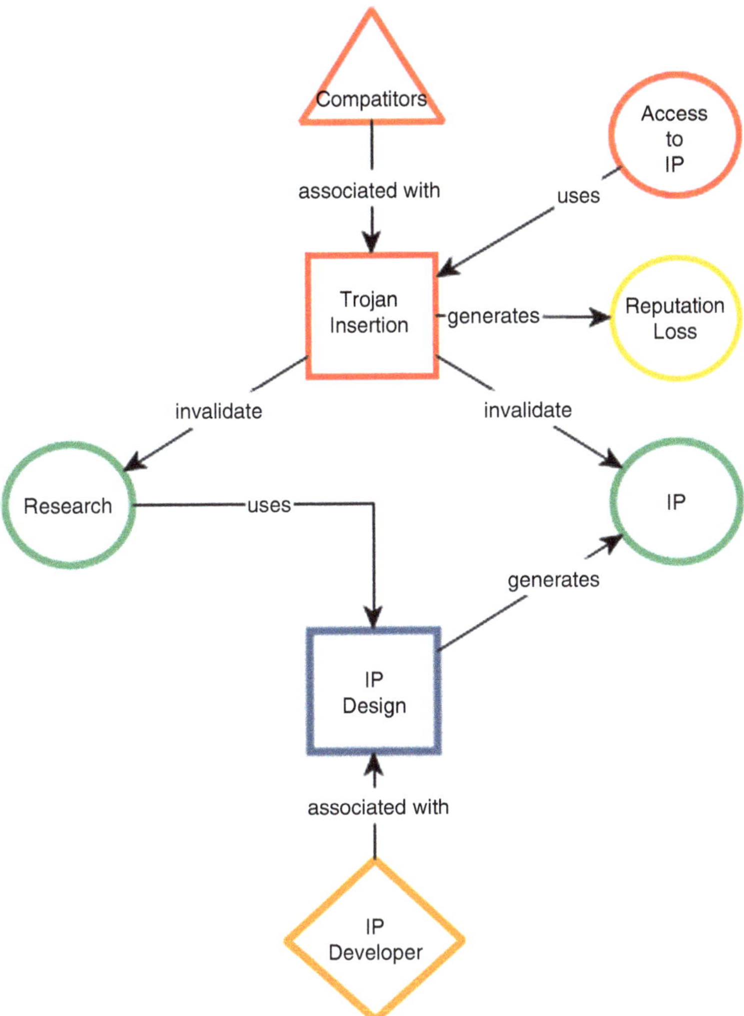


**Fig. 1.4** Threat modelling of hardware Trojan attacks on IP vendors

where the attackers can interface with the target), which should be enumerated at this stage of the threat modelling process. In this example, the entry points include a malicious insider in the design house, a malware in the circuit design tools, and vulnerability in the IT infrastructure. There are a number of potential attackers, in this case, which include, a competitor or a malicious tool developer, or even a state-funded institution. However, it is important to identify the class of the most likely adversary in order to develop appropriate countermeasures.

The important aspect of this method is that it provides a systematic approach to think about each threat, its motivation, mechanisms and consequences. Furthermore, it will allow developing a range of new attack scenarios by combining different mechanisms.

The same approach can be used to model different types of attacks on various stages of the IC supply chain as shown in the simplified example in Fig. 1.5, wherein an exemplar attack scenario is considered for each stage. The following section will provide a comprehensive review of current attacks, their preconditions, locations in the supply chain and potential impacts.

## 1.6 Identification of Threats

This section gives a comprehensive summary of possible attack mechanisms related to each of the adversary four main goals: counterfeiting, information leakage, sabotage and tampering. It is worth noting that some attack techniques discussed below can be used to achieve more than one goal.

### *1.6.1 Counterfeiting*

The aim of the attackers, in this case, is to fraudulently imitate an original IC to produce a counterfeit. The latter is usually an unauthorized copy that is relabelled, overproduced, cloned or recycled. If successful, such an attack undermines the "authenticity" property and leads to an IC with inconsistencies in the performance, reliability and functional characteristics. The counterfeiting threat is caused by the globalization of the semiconductor industry, which made it feasible for numerous, potentially malicious, parties to be involved in the design, fabrication, development and distribution of ICs. In fact, each of the component in an electronic system (e.g. embedded processors, analogue to digital converters, DSP circuity, sensors, printed circuit boards (PCBs)) is sourced from a diverse group of suppliers from around the globe. To give some perspectives on this, Apple uses more than 200 suppliers form over 40 countries to produce its iPhones [18]. The most common types of counterfeiting mechanisms are summarized below:

- Recycled ICs: wherein electronic components are collected from E-waste (such as discarded PCBs), then repackaged and sold in the market as new. Such devices can typically function as normal, but it will have reduced performance and reduced reliability due to the ageing of CMOS technology and the chip extraction process [19].
- Remarked Devices, wherein the marking on the package or the die is modified to include false information, for example electronic devices designed for commercial applications can be remarked with higher specifications so that it can be sold at a higher price to be integrated in industrial or military grade systems.

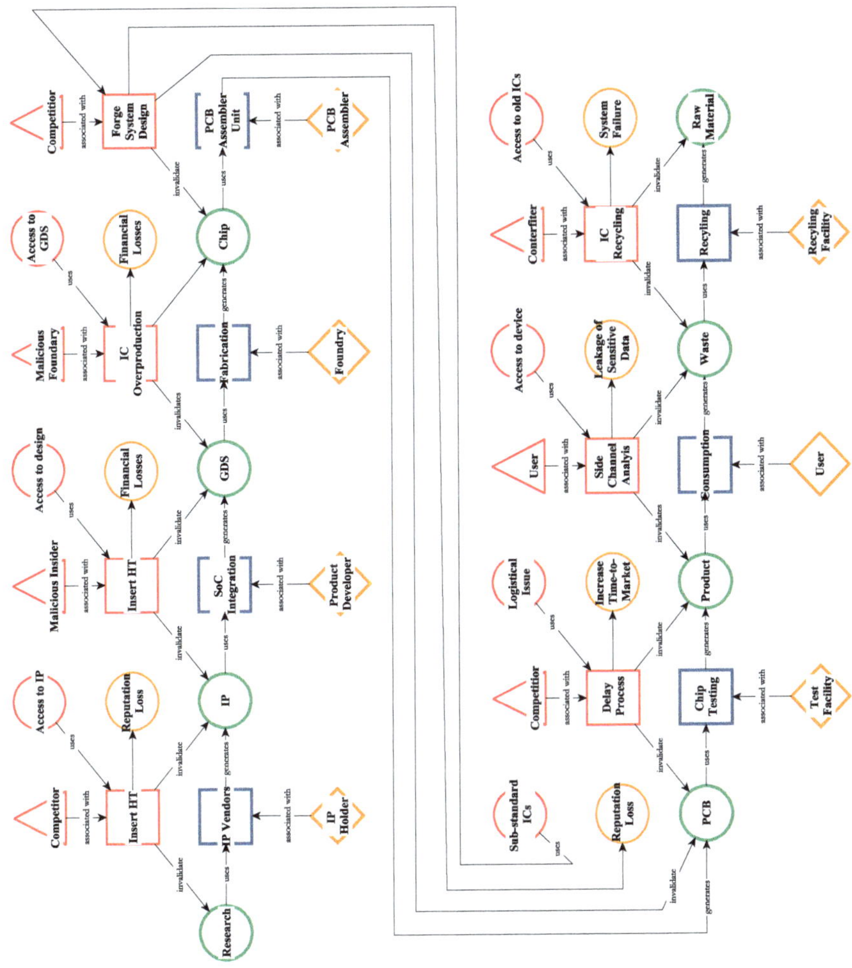


**Fig. 1.5** An Exemplar threat modelling of IC supply chain

- Defective ICs are chips that have failed the functional or parametric tests or found to be out of spec, and subsequently placed in the market as authentic products.
- Overproduced ICs: In this case, an untrusted foundry or rogue elements who have access to the original design produce more chips than the contracted amount, which are then sold illegally, at potentially lower price. These devices typically have reliability problems as they would not have undergone the same testing procedures as the authentically produced chips.

It has been estimated that at least 1% of semiconductor products are counterfeits, this percentage is expected to rise given the continuous drive for cheaper electronics and the increase of the sophistication and effectiveness of the tools and technologies used to produce counterfeit ICs [20]. The economic loss, on the UK only, is estimated to be 30 billion [1]. Furthermore, the global spread of this problem makes plausible for systems' developers to unknowingly utilize counterfeited components when building their products. Such a scenario may lead to catastrophic consequences if such devices are used in crucial sectors such as critical infrastructure, military applications, food and medicine. There already reported incidents of such cases, including malfunctioning military weapons and vehicles due to counterfeited parts [2], a loss in human lives (e.g. deaths due to contaminated food, such as this year *E. coli* infection).[3]

A summary of counterfeiting-related attacks, their preconditions, location in the supply chain and potential impact is provided in Table 1.5.

## 1.6.2 *Information Leakage*

Attackers aim to extract sensitive data from an IC at various stages of its implementation. The information that could be leaked may be related to the Intellectual properties (IP) of the design itself, secret cryptographic keys stored on chip or any other sensitive data associated with the hardware or the software downloaded to the IC either prior to or during operation.

The great variety of sensitive information that may exist in an electronic device means there are a wide range of attacks depending on the type of information to be targeted. This section classifies information leakage attacks into two types according to their goal: data theft and IP piracy, as explained below.

### 1.6.2.1 IP Piracy

This type of attack aims to steal design secrets and intellectual properties (IP); it can be performed in a number of ways, including, but not limited to, the following scenarios:

[3]https://www.cdc.gov/ecoli/2018/o157h7-04-18/index.html.

**Table 1.5** Threat modelling of counterfeiting

| Attack | Impact | Locations | Precondition | | Attack difficulty | Most likely attacker class(es) |
|---|---|---|---|---|---|---|
| | | | Assumed adversary capabilities | Root of vulnerability | | |
| IC overproduction | Financial losses.<br>Damage of reputation.<br>Safety risks due to untested chips. | Fabrication | Access to design files<br>Unrestricted access to a fabrication facility | Outsourcing the IC fabrication<br>Ease of access to IC black markets<br>Ineffective regulations or law enforcement measures to protect IPs<br>Technical difficulty associated with detection of overproduced chips | >3 | 3 |
| Defective chips | Financial losses<br>Damage of reputation<br>Safety risks due to untested chips | Testing | Access to testing facility | Ease of access to IC black markets | 1 | 3 |
| Remarking ICs | Financial losses<br>Damage of reputation<br>Safety risks due to untested chips | Fabrication<br>Testing<br>Packaging<br>User<br>Discarded chips | Access to fabricated chips<br>Access to IC remarking tools | Ease of access to fabricated ICs, through direct purchase or by obtaining discarded or old chips<br>Availability of remarking technologies<br>Unmatched demands for certain types of ICs (e.g. military grad, discontinued chips) | 2 | 3 |
| Recycling ICs | Financial losses<br>Damage of reputation<br>Safety risks due to untested chips | Discarded chips | Access to old ICs<br>Access to IC remarking tools | Availability of remarking technologies<br>Unmatched demands for certain types of ICs (e.g. military grad, discontinued chips) | 2 | 1, 3 |
| Cloning ICs | Financial losses<br>Damage of reputation<br>Safety risks due to untested chips | Fabrication<br>IP design<br>SoC integration<br>Testing<br>Packing<br>User<br>Discarded ICs | Ability to obtain a gate-level netlist of the chip through reverse engineering or other IP piracy methods<br>Access to a fabrication facility | Ease of access to IC black markets<br>Lack of regulations or law enforcement measures to protect IPs<br>Technical difficulty associated with detection of cloned chips | 3 | 3 |

(a) Cyberattacks on IP companies

In this case, hackers bypass the security defence mechanisms of the computing infrastructure of an IP design company and copy design secrets. The feasibility of such a scenario has been demonstrated in a number of attacks.[4] This scenario is considered in the literature to be the most economically significant form of IP piracy, hence requires more attention [21].

(b) IP theft by rogue designers

In this case, a malicious engineer in the SoC design house, who has access to third party IPs, can steal design secrets.

(c) IP theft at the fabrication stage

A rogue element in the IC foundry, who has access to the design files, and use these to extract the gate-level netlist and reverse engineering the design to infer its functionality.

(d) Reverse engineering

This mainly involves two steps. The first is *De-capsulation* that is the removal of the chip's packaging. The second is *De-processing* which consists of removing the chip layers one by one in reverse order and photographing each layer, these information will be used to re-construct the netlist and ultimately expose design secrets. Reverse engineering tools are becoming more sophisticated with the use of advance digital image processing techniques [22, 23] and machine learning algorithms [24, 25]. This trend combined with the ease with which silicon chips can be obtained, through direct purchase for example, make reverse engineering a rising threat. Pirated IPs are used to produce cloned chips, which is also considered a form of counterfeiting.

A summary of IP piracy-related attacks, their preconditions, location in the supply chain and potential impact is provided in Table 1.6.

#### 1.6.2.2 Data Theft

In this work, data theft refers to a group of security attacks that aim to exploit vulnerabilities in the design of the hardware system to obtain sensitive information such that encryption keys, private data and user credentials. This takes place after the chip is deployed in the field; therefore, the attacker is assumed to have access to the design and able to perform experiments. There are large number of attacks that fall under this category, these include, but not limited to, the following mechanisms.

(a) Microprobing

This attack consists of attaching a microscopic needle onto the internal wiring of a chip, which allows reading out internal signals and revealing sensitive data that are not meant to leave the chip. This attack requires de-capsulation and

[4] https://www.ipwatchdog.com/2020/03/10/global-threat-report-key-takeaways-ip-intensive-companies/id=119705/.

**Table 1.6** Threat modelling of IP piracy attacks

| Attack | Impact | Location(s) | Preconditions | | Attack difficulty | Most likely attacker class (es) |
|---|---|---|---|---|---|---|
| | | | Assumed adversary capabilities | Root of vulnerability | | |
| Cyberattacks on IP companies | Financial losses | IP vendor | Access to the computing infrastructures of the IP developers | Weak security defences in the IT infrastructures | 2 | 3, 4 |
| IP theft | Financial losses | SoC design | Access to the computing infrastructures of the IP developers | | 2 | 1, 3, 4 |
| Reverse engineering design files | Financial losses | IC fabrication | • Access to the design files<br>• Able to extract the IP netlist using a range of tools and reverse engineering technologies | Correlation between circuit layout and the gate-level netlist and ultimately the design functionality | 3 | 3, 4 |
| Chip reverse engineering | Financial losses | Testing<br>Packaging<br>In the field | • Access to a working chip files<br>• Able to extract the IP gate-level netlist using a range of tools and reverse engineering technologies | Correlation between circuit layout and the gate-level netlist and ultimately the design functionality | 4 | 3, 4 |

de-processing of the chip similar to the reverse engineering process discussed above.

(b) Side-channel analysis

These attacks exploit the dependency between the secret information and the physical characteristics of the system when such data are computed, such as power consumption [26], computation time [27] and electromagnetic emission [28], of a combination thereof. One embodiment of this type of attacks is differential power analysis [26], which aims to deduce the secret keys from a smartcard by analysing the power measurements of its chip.

(c) Speculative execution attacks

This class of attacks exploit the weaknesses in the hardware architecture that results in unintentional leakage of sensitive data. A prime example of this type is the Spectre attack [29]. The latter breaks the isolation between different applications running on the same machine, which makes it feasible for a malicious process to steal/copy sensitive data from a victim process. Spectre affects processors that use branch prediction and speculative execution to maximize performance, this is because, in some cases, the speculative execution following a branch miss-prediction can reveal private data to attackers. For example, if this speculative execution performs a number of memory accesses, such as the pattern of these operation is dependent on secret data, then the resulting state of data cache can leak information through which an attacker may be able to deduce the private data using timing attacks. Another example of architectural attack is Meltdown [30], which breaks the isolation between user applications and the operating system, this effectively makes it feasible for a malicious programme to have full access to the memory and the secrets stored by other software running on the same machine or by the operating system. Meltdown exploits an inherent vulnerability in many processors, which allows a malicious process to bypass the privilege checks that prevent it from accessing data belonging to the operating system and other running processes. Both Spectre and Meltdown attacks assumes the adversary has access to the computing device under attacks and able to install their software, they are also assumed to have advanced knowledge of the underlying system.

Other speculative execution attacks have also emerged such as the *Foreshadow*, which allows virtual machines (VMs) to read the physical memory belonging to other VMs or the hypervisor [31].

All above-mentioned attacks require additional steps to retrieve the targeted sensitive data, this is typically archived using cache side-channel attacks, which will be discussed later [32].

(d) Cache timing attacks

This attack uses timing information related to cache memory access operations to reveal sensitive information such as encryption keys.

The ever-increasing performance gap between the processor and memory technologies, respectively, has led to designer to introduce smaller pieces of memories, referred to as Caches, between the CPU and the main memory. These are placed in close physical proximity to the CPU and used to store recently used data. This greatly reduces memory access time, hence enhancing system performance. However, the use of caches introduces a security vulnerability because of the time variation between a cache hit and a cache miss. The former is the time required to access a data item present in the cache, whereas the latter is the time taken to retrieve a data item not present in the cache. This time difference is very large and can lead to measurable performance difference. Typically, if the content of specific memory address is cashed, the next access to the same or adjacent memory addresses that are mapped to the same cache line will be quick. Otherwise, if the address is not in the cache, next access will be slow.

Cache timing attacks exploit this timing variations to deduce the memory access patterns of the targeted victim process, which can reveal secret information [33]. There are a number of methods to carry out this type of attacks, all of which assume the adversary is able to install a malicious process into the target computing device and measure the execution time of a number of tasks.

One of the first realization of this attack is based on the "Evict + Time" approach [33]. The latter requires the aggressor process to trigger the victim process, evict its cached data, and then measure the execution time of the victim process. To explain how such an attack can reveal secret information, let us consider an AES encryption to be the victim process. When this is triggered for the first time, some data from the AES lookup tables will be loaded into the cache. If the aggressor process manages to evict these data from the cache, the execution time of the encryption process will be longer when it runs for the second time as these data have to be fetched from the main memory. Otherwise, if the aggressor process fails to evict the lookup tables data, then encryption will be faster in the second run. This means the length of the execution time of the second run of the victim process reveals its memory access patterns. In the case of AES encryption, wherein the indexes of lookup tables are computed by the private key, these timing information reveal part of the key. Other variations of this approach include the "Prime + Probe" [33], in this case, the attacker only measures its own execution time rather than that of the victim process, which makes it more effective and noise-resistant compared to the "Evict + Time" approach. A third technique termed "Flush + Reload" [34], which gives the attacker more control on the data to be evicted from the cache. This is achieved through the use of "cflush" instruction that can evict specific memory lines from all levels of the cache, which allow this approach to achieve more accuracy.

(e) Memory attacks

There are a variety of mechanisms that can be used to extract sensitive information from memory blocks. These ranges from software-level attacks to invasive physical attacks. Examples of software-based mechanisms include snooping and read-out attacks [35], which involves the use of malware. Examples of hardware-level attack includes data remanence attacks on flash memories [36] and SRAM [37], these exploit the correlation between the data stored and some of the electrical characteristics of the target memory cell. For example, the shift of threshold voltage of the SRAM transistors due to BTI ageing is correlated to the data being stored [38].

(f) PUF cloning attacks

Physically unclonable function technology has been developed as a more secure alternative to typical memory technologies. It allows a hardware-based identity by exploiting the intrinsic variability of the IC manufacturing process. PUF technology has now a wide range of security applications, which include, but not limited to, cryptographic key generation, authentication schemes, hardware metering and secure lightweight sensing [39]. There are two types of PUF constructions, weak and strong. The former typically have smaller number of challenge/response pairs (i.e. input/output); in other words, the total size of their output space is smaller than that of strong PUF. Cloning attacks aim to produce a physical or a mathematical clone of this primitive. A number of physical cloning attacks on SRAM and Arbiter PUF design have been reported [40].

PUF modelling attacks aim to construct a mathematical clone of the PUF, which can be used to mimic its behaviour to a high degree of accuracy, typically more than 99.9% [40]. This attack, which targets strong PUFs, is performed in two steps. First, the adversary collects a large number of challenge/response pairs (CRPs) by eavesdropping on a communication channel where these CRPs are transmitted or by physically probing the input of the PUF, applying a polynomial number of challenges and collecting the respective responses. The method with which these CRPs are collected is very much dependent on the assumed capabilities of the adversary in question. Second, the attacker will feed these data into machine learning algorithms, such as Logistic Regression, Evolution Strategies or artificial neural network, which generate a software model of the PUF. Several known Strong PUF classes have been attacked, including Arbiter PUFs, XOR Arbiter PUFs, Feed-Forward Arbiter PUFs. This attack severely undermines the security benefits of the PUF technology as it invalidates its most fundamental assumption, i.e. the unclonability of PUF.

A summary of data theft-related attacks, their preconditions, location in the supply chain and potential impact is provided in Table 1.7.

### *1.6.3 Sabotage*

Attackers aim to deliberately damage or destroy an electronic device such that it becomes lost, unavailable or unusable. This attack can take place at any stage of the life cycle of the IC. One possible attack mechanism is causing deliberate delay to one or more of the IC production processes to increase the time to market, hence gives an undue advantage to competitors. Other examples are intentional erasure of programme or data, denial-of-service attack, physical destruction of hardware devices. There are large number of mechanisms that fall under this category, these include, but not limited to, the following examples.

**Table 1.7** Threat modelling of data theft attacks

| Attack | Impact | Location(s) | Preconditions | | Attack difficulty | Most likely attacker class(s) |
|---|---|---|---|---|---|---|
| | | | Assumed adversary capabilities | Root of vulnerability | | |
| PUF modelling attacks | Unauthorized access to resources | User | Access to PUF response/challenge pairs | PUF design is simple hence can be modelled using machine learning algorithms | 2 | 2 |
| Side-channel analysis | Loss of sensitive data | User | Access to the computing devices<br>Able to perform non-invasive experiments (e.g. measurement of power consumption, execution time or electromagnetic emissions) | Correlation between side-channel information and secret data being computed | 3 | 3, 4 |
| Speculative execution attacks | Loss of sensitive data<br>Reputation damage of IP providers | User | Advanced knowledge of computer architectures<br>Access to the computing devices to measure execution times of various running processes | Speculative operations can affect the micro-architectural state, such as information stored in Translation Lookaside Buffers (TLBs) and caches, which may lead to leakage of sensitive data when combined with Cache side-channel attacks | 3 | 2, 3 |
| Microprobing | Loss of sensitive data<br>Destruction of the chip | User | Access of reverse engineering tools<br>Knowledge of IC implementation<br>Physical access to the device | The transmission of sensitive information on the internal wires without sufficient protection | 4 | 3, 4 |
| Cache timing attacks | Loss of sensitive data | User | Advanced knowledge of computer architectures<br>Access to the computing devices to install a malware and measure execution times of various running processes | The dependency of the memory access time on the location of data item being fetched (e.g. whether or not it is present in the cache or the main memory) | 2 | 3 |

#### 1.6.3.1 Cyber-Physical Attacks

These attacks exploit vulnerabilities in the software or firmware to disable or compromise computing devices. These can be used to facilitate criminal activities by disability CCTV cameras or to disrupt emergency services response by causing artificial traffic jam and interfering with local communications. One of the most widely publicized cyber-physical attack is Stuxnet. The latter is considered by many as a game changer as it shows the possibility of exploiting software vulnerabilities to inflict physical damage on industrial control systems [41]. The latter have previously been considered immune to cyberattacks due to their perceived insolation. Stuxnet was a 500-kB computer worm that targeted the software of more than 14 industrial sites in Iran, including a uranium-enrichment plant, which is believed to have caused substantial damage to the nuclear programme. Unlike a typical computer virus, a worm is capable of replicating itself in order to spread to other computing devices and systems on the same network. Stuxnet is thought to have spread with the aid of USB thumb drives physically inserted into the control machines.

Since the discovery of Stuxnet, a number of attacks have emerged that are based on the same principles, i.e. leveraging software vulnerabilities to introduce a computer worm that facilitate sabotaging the target physical systems. Examples of these include Triton that was first discovered at a petrochemical plan in Saudi Arabia, it is capable disabling industrial safety systems. Another example is CRASHOVERRIDE, which caused the Ukraine malware power outage. A comprehensive treatment of this type of attacks can be found in [42].

#### 1.6.3.2 Rowhammer Attacks

This is a form of fault attack which exploits the fact that repeated accesses to DRAM rows can cause bits to flip in adjacent DRAM rows [43]. There are a number of physical mechanisms that have been suggested to cause the rowhammer errors, including the electromagnetic coupling between adjacent DRAM memory cells, charge leakage and hot carrier injection.

Rowhammer attack can be used as a sabotage mechanism by waging a persistent attack to cause large number of errors. What makes this type of attacks particularly worrying is that it can be waged remotely by sending network packets as has been demonstrated in [44]; furthermore, it does not require expensive resources.

Additionally, Rowhammer techniques can be used to undermine the integrity of electronics systems by facilitating an elevation of privilege attack, which allows a potential adversary to gain control of the system and bypass security mechanisms. A number of studies have demonstrated how this can be achieved in practice including its use to gain kernel privileges on x86-64 Linux [45] and on mobile platforms [46].

#### 1.6.3.3 CLKscrew Attacks

This is a form of fault attacks that leverages the software to modify physical properties of the hardware. It exploits the energy management features to induce errors, for example by adjusting the configuration of Dynamic Voltage and Frequency Scaling (DVFS) features beyond normally allowed operating points [47]. This attack, which does not need a physical access to the system, can be used as a sabotage mechanism to crash a system by causing too many errors. Additionally, it can be employed to leak-sensitive data and for escalation of privileges as has been demonstrated on android devices in [47].

#### 1.6.3.4 Attacks on IC Production

This is another form of sabotage, wherein an adversary attempts to hinder the IC production process in order to delay the release of a new product (e.g. a new smart phone). In this case, the attacker may target transport links, fabrication facilities or testing sites. Such attacks can be performed physically or through the cyber space (e.g. a cyberattack on the chip testing equipment).

A summary of sabotage-related attacks, their preconditions, location in the supply chain and potential impact is provided in Table 1.8.

### *1.6.4 Tampering*

The goal of the attacker is to undermine the integrity of a system and its ability to function as expected. Unlike sabotage, adversaries, in this case, do not want to completely destroy the system, their aim is to trigger faulty behaviours in electronic devices, which helps reveal secret information (e.g. encryption key) that are hard to obtain otherwise. This threat can affect the IC at any stage of its life cycle. A prime example of these attacks is injecting a fault to an operational system through electromagnetic radiation or voltage glitches. Another example is Trojan insertion, which consists of manipulating the design files to introduce an extra malicious functionality or facilitate the leakage of sensitive information. There are a great many examples of attacks that fall under the tampering category, and these include, but not limited to, the following mechanisms.

#### 1.6.4.1 Fault Injection Attacks

In this case, an adversary can induce errors during the computation of a cryptographic algorithm to generate faulty results, which can subsequently be exploited to deduce information about the secret key stored in the electronic device. Fault attacks are more effective in breaking unprotected systems compared to side-channel

**Table 1.8** Threat modelling of sabotage attacks

| Attack | Impact | Location(s) | Preconditions | | Attack difficulty | Most likely attacker class(s) |
|---|---|---|---|---|---|---|
| | | | Assumed adversary capabilities | Root of vulnerability | | |
| Stuxnet-type attacks | Physical damage to system | User | Ability to install a computer worm into the machine (either remotely or physically)<br>Sophisticated knowledge of the target system | Software | 5 | 3, 4 |
| Rowhammer | System failure<br>Bypass of security mechanisms | User | Access to the computing devices<br>Able to perform non-invasive experiments (e.g. measurement of power consumption, execution time or electromagnetic emissions) | DRAM physical structure and fabrication technology | 2 | 1, 2, 3, 4 |
| Clkscrea | System failure<br>Bypass of security mechanisms<br>Leakage of secret information | User | Advanced knowledge of computer architectures<br>Access to the computing devices | Unfettered software access to energy management hardware<br>Ability of the hardware regulators to be able to push voltage/frequency past the operating limits.<br>Using the same power domain across security boundaries | 3 | 1, 2, 3 |
| Attacks on IC production facili-ties/processes | Increase time to market<br>Financial loss<br>Reputation damage | User | Knowledge of IC production procedure<br>Access to infrastructure | Insufficient protection of IC manufacturing sites, testing facilities, packaging centres or transport links | 3 | 3 |

analysis. For example, an attacker can break an AES encryption with two faulty results compared to the large amount of measurements required using differential power analysis. This type of attacks, however, does require prior knowledge about the design, in order to choose the locations of the injection, the number of faults required, and the most appropriate injection mechanisms. There are a number of fault injection techniques, as summarized below.

(a) Electromagnetic radiation
This technique consists of generation of a sudden fluctuation of the magnetic field in neighbourhood of the target chip surface. This variation will potentially induce parasitic currents, which disrupt the operation of the target device. An important advantage of this approach is that it does not require de-packaging the IC and can be performed relatively cheaply (e.g. using a needle wound with wire).
(b) Laser attack
This technique exploits the photoelectric effects by applying a laser beam to a specific area of a device to induce a parasitic current in the electronic circuitry. This can potentially cause a computation error, leading to a faulty output. The success of this approach relies on the attacker's ability to appropriately control the beam's energy, polarization, target location and emission time.
(c) Temperature attack
This approach consists of running an electronic device outside the range of its operational temperature. The use of extensive heating as fault injection technique has been experimentally demonstrated in [48], wherein the author showed how to successfully compromise the security of an RSA decryption. Memory blocks are also vulnerable to this type of attacks as extreme temperature can induce errors in both volatile and non-volatile implementations [49]. This attack can also be performed relatively cheaply, assuming the adversary has sufficient knowledge of the system under attack.
(d) Glitch attacks
This attack consists of inducing variations in the supply voltage or in the external clock pin to impede the functionality of the device. The success of this approach relies on the attacker's ability to appropriately control the amplitude and the duration of the glitches. This approach allows the fault to propagate across the device, therefore focusing on specific location is not possible. On the other hand, it allows the adversary to induce errors in rarely activated nodes. This is probably the most common form of fault injection techniques due to the ease of its application mechanism.

#### 1.6.4.2 Software-on Hardware Attacks

This is a group of attacks that leverage the software to get the hardware to behave in an erroneous manner. Examples of this type include Stuxnet, Rowhammer and

Clkscrew attacks, which are used for sabotage (i.e. to compromise the system's availability) and can also be used for tampering (i.e. to compromise the system's integrity).

#### 1.6.4.3 Trojan Insertion

A hardware Trojan (HT) is a malicious addition or modification to the existing circuit elements, in order to change the system functionality, reduce its reliability or leak valuable information. HT can be inserted at different stages of the lifecycle of a system and at different abstraction layers, for example, in the RTL code or netlist of an IP (Intellectual Property module) during design phase, at the integration process of a SoC, or in the layout of an integrated circuit during manufacturing. HTs have typically two main components: the trigger circuit that is used to activate the malicious operation, and the payload which is the circuit that performs the malicious operation. Although, this attack can take place at any stage during the development of an electronic system, it can be most effective if performed in the early phases of the design process (i.e. IP development or SoC integration). Hardware Trojans can be classified depending on their physical characteristics, activation mode, the action that they perform, and the attack target. In terms of physical characteristics, Trojans can be classified as analogue [50], digital or hybrid [51, 52]; in all cases most of the Trojans modify the structure of the system, while others only change parametric characteristics like the doping concentration of transistors [53, 54]. In terms of activation method, Trojans can be triggered externally using inserted antennas and sensors [55], or by changing the circuit operating conditions like the clock speed [54] or supply voltage [53]. Trojans can also be activated internally when a certain logic condition is met or remain always activated. In terms of the action, Trojans can have a functional effect, for example to disrupt services and leak information, or a parametric effect, for example, degrading the performance of a system or reducing its reliability [56]. Finally, Trojans can be classified according to the attack target either by the type of technology where the Trojan is inserted, FPGA or ASIC, or by the design phase at which the Trojan is inserted, for example, during the design of an IP by inserting the Trojan in the RTL code, during the system integration of an SoC, or during the manufacturing of the Integrated Circuit (IC) at layout level by modifying the GDSII files. There are a great many examples of how hardware Trojan can be exploited to undermine the security of electronics devices by facilitating the leakage of sensitive information. For example, the work in [57] describes a Trojan that facilitates a Differential Fault Analysis (DFA) using Piret's algorithm to find the key of a hardware implementation of AES by evaluating only two encryptions. The Trojan payload consists of one XOR that inserts a fault in one bit of the state in the eighth round; the trigger circuit consists of AND gates that tests a specific input combination of the input and that the round counter is equal to 8. A second example is the work in [58], which demonstrates the design of a Trojan to perform an attack of an AES encryption system and a method to obtain the key using the outputs of the cypher of additional encryption rounds The trigger circuit uses signals of the test mode to decrease the possibility of being detected, while the payload circuit allows reading the output of subsequent rounds of encryption after the tenth round which are used to obtain the key with the proposed method. The work in [59] describes how by a reverse engineering process a Trojan can be inserted in the bitstream of an FPGA design to bypass power-up self-tests in order to manipulate other components of the system to perform fault attacks without being detected. They also present a semi-automatic method to detect the S-Box implementation in the bitstream and leak information from the first round via an inserted UART/RS-232 module. Then the adversary can compute the key by applying the inverse S-Box operation followed by an XOR with the plaintext. Trojans can also be used to facilitate the implementation of side-channel attacks or to circumvent the protection against them. Lin et al. [60] present a technique called MOLES to perform power side-channel attacks by inserting a circuit that consumes power depending on data. On the other side, Ender et al. [4] present a technique to perform power side-channel attacks on masked implementations of cyphers by breaking their uniformity property.

A summary of Tampering-related attacks, their preconditions, location in the supply chain and potential impact is provided in Table 1.9.

## 1.7 Mitigation Techniques

This section provides a comprehensive review of the attack mechanisms discussed above.

### *1.7.1 Countermeasures for Counterfeiting Attacks*

The defence mechanisms against this type of threat can be classified, according to their goals, into two types, detection techniques and prevention methods, as will be detailed below:

#### 1.7.1.1 Counterfeit Detection Techniques

This type of countermeasures makes it feasible to differentiate between an authentic and a forged device, by comparing their respective electrical or physical characteristics, or by embedding unique identifiers at the circuit level. This area of research is well developed with some solutions already deployed in commercial products.

**Table 1.9** Threat modelling of tampering attacks

| Attack | Impact | Location (s) | Precondition | | Attack difficulty | Most likely attacker class(es) |
|---|---|---|---|---|---|---|
| | | | Assumed adversary capabilities | Root of vulnerability | | |
| Fault injection attacks | Leakage of secret information<br>Bypassing security mechanisms<br>System malfunction | User | Knowledge of the system architecture and functionality<br>Access the hardware<br>Able to perform semi-invasive experiments | Susceptibility of electronics circuit to temperature variations, supply voltage fluctuations and electromagnetic interference | >1 | 1, 2, 3 |
| Hardware Trojan | System failure<br>Bypass of security mechanisms<br>Leakage of sensitive information | IP vendors<br>SoC<br>Integration<br>Fabrication | Access to design files (e.g. RTL, GDS, ...)<br>Access to design tools<br>Sophisticated knowledge of IC design | Outsourcing of IP development and IC fabrication<br>High complexity of integrated circuits that makes it harder to detect Trojan | >3 | 3, 4 |

#### Counterfeit Detection: Physical Inspection

Counterfeit devices, in particular, recycled ICs, have different physical properties compared to authentic chips, for example, scratches on the package or deformed leads. These differences can be spotted by visual inspection using magnification lamps. More advanced techniques include the use of X-ray imaging of the die or the bond wires, which can be performed without de-packaging. Chemical composition analysis through spectroscopy is another approach that have been reported [20].

#### Counterfeit Detection: Side-Channel Analysis

This approach relies on measuring the electronics properties of the suspected components (e.g. performance, power consumption, electromagnetic noise), it assumes that these metrics are going to be different compared to those of the authentic devices [61].

For example, the authors of [62] have proposed an approach to detect recycled FPGAs based on exhaustive characterizing the path delay of the lookup tables (LUT), wherein machine learning algorithms are also used to improve the efficiency of the detection. A similar approach, presented in [63], relies on the extraction of an electromagnetic signature to detect both recycled and cloned ICs. A third technique [64] is based on estimating a spatially integrated electromagnetic signature of a given IC, by measuring its complex reflection coefficient when illuminated with an open-ended rectangular waveguide probe, at the microwave frequencies K-band (18–26.5 GHz) and Ka-band (26.5–40 GHz), respectively. This signature is a function of the IC internal material properties, geometry and metallic deposition of circuit element, and wire bonds, which means aged (including recycled) ICs exhibit markedly different reflection properties than their reference and new counterparts.

The above techniques need access to data taken from golden chips, which can limit their practical use. There are, however, emerging techniques that do not need golden data. For example, the authors of [65] have demonstrated the possibility of detection of recycled ICs by measuring the dynamic current ($I_{DDT}$). The key idea of this technique is that different parts of a digital block have different stress levels due to varying level of activities. For example, the data path units, such as an adder, may experience different activities in their upper and lower due to the abundance of narrow-width operands, this means the circuitry in these two different parts will age at different rates. This makes is feasible to use the correlation of ($I_{DDT}$) of these two parts to identify unbalanced shift in the threshold voltage $V_{th}$ due to aging. There are two undelaying assumptions in this case: (1) this unbalance only exist in recycled chips, (2) it can be differentiated from the expected unbalances in authentic chips, caused by the manufacturing process.

Overall, the effectiveness of these techniques, especially those relying on aging effects [63, 65–67] is dependent on the amount of shift of the electrical characteristics due to prior usage.

Counterfeit Detection: Fingerprinting

The essence of this approach is to embed a unique identifier (i.e. a fingerprint) in each IC, which makes it distinguishable from other ICs with an identical functionality, even IF produced by the same factory using the same fabrication technology. This makes it feasible to trace each IC throughout its life cycle and allows detection of counterfeit components.

There are a number of approaches to achieve this, which include the following:

1. Conventional serial numbers

   These are physically indented on the devices or stored in an on-chip memory. Extrinsic tags are easily reproducible, hence are no longer effective given the increase in the sophistication of counterfeiting techniques.
2. DNA marking

   This approach consists of embedding a tag on chips and/or circuit boards, made from botanical DNA sequences [68]. Special chemical, called fluorophores, which glow under specific wavelengths of light, are used as part of the tagging process. To verify the authenticity of an electronic component, one will need to scan in the fluorescent signature to check it exists. Following this, a sample of that tag is swabbed to be checked using standard forensic DNA techniques. The scanned sequence is then checked against the product's database to confirm whether the label and the part matches. This approach provides a cloning-proof solution for product tagging, which is a great deal more secure compared to conventional serial numbers; however, this approach has prohibitive costs, in term of the price of each tag and the time it takes to verify it [69]. This limitation can greatly limit the range of electronic products that can use this technology to defend against counterfeiting.
3. Physically unclonable functions

   These security primitives exploit the intrinsic variations of the IC manufacturing process to generate a unique digital identity for each chip. Although, this technology has received a great deal of attention since it was first discovered in the seminal work in [70], its adoption commercially for device authentication has been rather slow. This is mainly due to two main outstanding problems:

   (a) The instability of the PUF response in the presence of noise, ageing and variability of operating conditions (such as supply voltage and temperature) [71].

   (b) The vulnerability of existing PUF architectures to a number of security attacks [40, 72], both to physically clone the PUF or to develop a software model, which has almost identical behaviour using machine learning algorithms.

   However, solutions to these challenges are emerging rapidly, which promises a greater adoption of PUF technology to mitigate against IC counterfeit. To improve the stability of PUF response, proposed methods include the use of error correction codes, temporal voting mechanisms and ageing acceleration techniques [40], other solutions include the use of self-correcting structure [73] which is claimed to have lower implementation costs. Countermeasures for PUF modelling attacks will be discussed in more details later on.

4. Digital fingerprinting

   The aim of this approach is to assign a unique and invisible identifier to each sold instance of the intellectual property (IP), this takes place during high-level, logic or physical synthesis. Compared to watermarking, digital fingerprinting allows the tracking of each individual sold IP. For example, if a designer suspects that their IP has been stolen and used in a counterfeit IC, they can check this by retrieving the embedded watermark, and then use the fingerprint to identify the IP buyer that has assisted (knowingly or unknowingly) the IP piracy. The main challenge when applying this approach to find an effective way of generating a unique fingerprint to each individual IP, while keeping the design overheads at minimum and reducing the time and complexity of the fingerprint generation process. The authors of [74] pioneered a digital fingerprinting approach, based on the optimization heuristics of VLSI design tools. This was achieved by exploiting partial solution reuse and the incremental application of iterative optimizers of a number of VLSI design-related NP-hard problems, such as partitioning, Boolean satisfiability (SAT), graph colouring and standard-cell placement, This allows specific set of constraints to be introduced in each iteration to create many unique fingerprints. A major problem in this type of digital fingerprinting is the significant increase in the design time and cost they may cause as, for example if the fingerprint is embedded at the layout level, each IC must have a different set of masks, which is simply not possible.

   More practical digital fingerprinting techniques [75, 76] have since emerged, which can be applied at the post fabrication stage. These are based on exploiting the observability do not care (ODC) and satisfiability do not care (SDC) conditions in logic design, respectively. For example, replacing the "And" gate in Fig. 1.6a with an "XOR" gate in Fig. 1.6b does not change the functionality because the nodes A and B can never assume a "00" value regardless of the value applied to the inputs X and Y. This is referred to as SDC in logic design. This approach can be used to give each chip a unique fingerprint by configuring such logic Gate, which can take place after fabrication.

#### 1.7.1.2 Counterfeit Preventing Techniques

This type of countermeasures aims to reduce the risk or completely prevent IC counterfeiting, with a combination of countermeasures at the circuit, systems and protocol levels. This is an active area of research with many emerging solutions, but these are yet to see wide adoption in the semiconductor industry, mainly due the cost associated with their use.

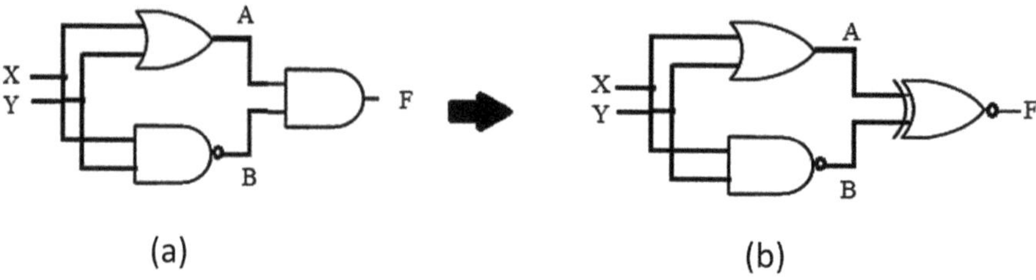


**Fig. 1.6** Digital fingerprinting for creating equivalent IPs. Circuits (**a**) and (**b**) have the same functionality by differs in one logic gate due to SDC

(a) Active IC metering

The aim of this approach is to prevent a malicious foundry from overproducing chips without prior agreement with the design house. This is achieved by embedding a locking mechanism into each fabricated chip, which renders it non-functional. Each IC has its unique key to unlock its functionality, which is provisioned by the deign house. There are two main types of active IC metering internal and external. The former method embeds the locking circuitry within the structure of the computation model in hardware design, in the form of a finite state machine (FSM). Consequently, this requires the inclusion of additional number of states and transitions, for example the number of required additional states should be no less than that of manufactured chips according to the approach presented in [77]. This increase in the design complexity is prohibitive, and may lead to unacceptable performance degradation that is hard to control, therefore subsequent work in [78, 79] aimed to reduce the overheads of this approach. External active IC metering implements the locking mechanisms at physical design stage and introduces cryptographic modules to protect the unlocking key. This technique was first proposed in [80], it consists of adding XOR gates on selected non-critical combinational paths, these gates are controlled by signals connected to the key register. The inclusion of these XOR gates renders the design non-functional unless the correct key is supplied; this approach, referred to as EPIC, is proven to be resilient against many of the known attacks. However, the addition of the XOR gate can degrade the timing of combinational paths, a problem that can be resolved using a slightly different locking scheme that controls the working mode of some specific scan cells in scan chain, which does not affect the timing of the functional path [81]. Another issue which affects the security of EPIC-type locking schemes is the need to unlock the circuit before the chip test, most probably at the fabrication facility, which makes the circuit vulnerable to a number of attacks [82, 83]. Countermeasures against such a threat include the use of a multiplexor-based locking strategy that preserves test response allowing IC testing by an untrusted party before activation [82]. It is worth pointing out that both external and internal active metering techniques, which rely on the use of PUF technology, are facing the same challenges, related to the stability of the PUF response, which have been discussed previously.

(b) Anti-fuse-based package-level defence

This approach aims to prevent two forms of IC counterfeiting, recycling and cloning; this is achieved through inserting a one one-time programmable anti-fuses (A) to selected pins inside the IC package [84], which disables their functionality. The locked pins need to be programmed before first-time use to make a chip functional in a system. In addition, the intrinsic random variations in programmed resistances of AFs connected to some of the remaining IC pins are exploited to create unique chip-specific signatures for authentication, which makes it feasible to detect cloned ICs. The advantage of this approach is that it does not require die-level design modifications, which reduces the overheads associated with IC locking schemes (i.e. hardware overhead, performance degradation and test cost); in addition, it can provide effective protection, even in the case of small chips with low pin counts such as analogue ICs, this is achieved at an 0.05 increase in the package area.

(c) IC supply chain assurance

This type of approaches aims to prevent or reduce counterfeiting incidents by enhancing the integrity of the hardware supply chain. A number of methods have been proposed in this direction. The work in [85] proposes an approach based on the use of consortium blockchain and smart contract technologies, wherein each chip in the supply is embedded with PUF for unique identification. This technique makes it feasible to trace ICs throughout their life cycle, even after they have been discarded. This is achieved through the storage of IC identification data in a decentralized and highly available chain, across multiple sites. A similar approach was also proposed in [86], using blockchain-based certificate authority framework, which manages critical chip information (i.e. identification, chip grade and transaction time). Additionally, there are a number of industry-led initiatives towards securing the IC supply chain [87, 88], most notably the IPC-1782 traceability standard, which created a single expandable and extendable data structure that can be adopted for all levels of traceability and enable easily exchanged information, as appropriate, across many industries. This standard can be applied to manual processes, as well as, fully automated processes and everything in between, which makes suitable for managing tractability data in the IC supply chain.

### *1.7.2 Countermeasures for Information Leakage Attacks*

The great variety of sensitive information that may exist in an electronic device (e.g. cryptographic keys, private data, IC design secrets) means there are a wide range of attacks depending on the type of information to be targeted. This section classifies information leakage attacks into two types according to their goal: data theft and IP piracy, and present countermeasures, accordingly.

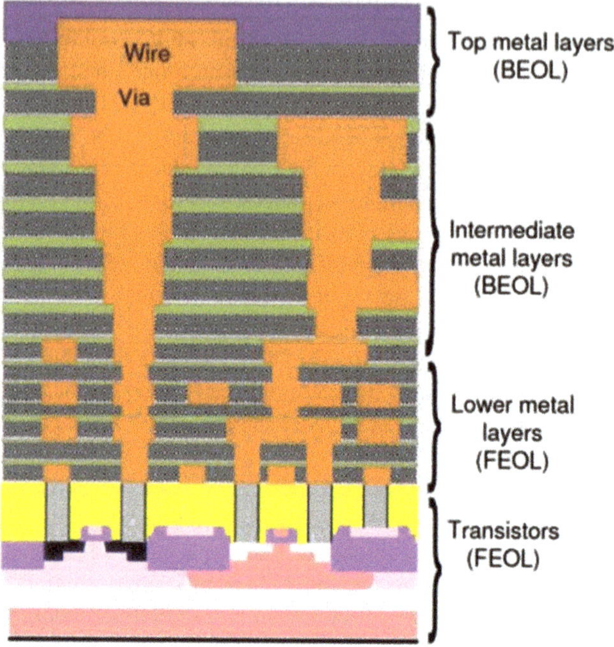


**Fig. 1.7** A cross section of a silicon chip

#### 1.7.2.1 Countermeasure for IP Piracy

Split Manufacturing

This technique was proposed to protect against IP piracy that is performed by untrusted foundries [89]. The latter have access to the layout files of the design, making it easy to recover the IP by rogue elements within the fabrication factory; therefore, the main goal of this technique is to restrict or reduce the amount of layout information given to potentially malicious parties. This is achieved by splitting the layout into two parts, FEOL (Front End of Line) and BEOL (Back End of Line), as shown in Fig. 1.7. The FEOL contains the transistors level and low-level metal layers, which requires advanced fabrication technology. In terms of the BEOL, it contains most of the remaining metal interconnect layers, which does not require expensive fabrication methods.

After fabrication, the wafers of these two parts will be combined using electrical and optical alignment techniques. This approach helps to hide the overall functionality of the design from an attacker residing at the FEOL foundry, thereby hindering them from pirating the design or maliciously modifying it via hardware Trojans.

However, the effectiveness of this defence mechanism is still far from established, and there are a number of reported attacks that can still deduce the full design based on the FEOL part only.

The root cause of this vulnerability is that the majority of commercially available physical design tools apply certain heuristics to minimize power, performance, and area, which may leak certain information. For example, to optimize performance, place and route tools place the blocks to be connected in close proximity, hence reducing the delay of the interconnect, this can be exploited by attackers to deduce the BEOL information by connecting each pin to the nearest neighbours. This type of attacks is typically referred to as a proximity attack. An example of these attacks was reported in [90, 91] and shown to render most prior protection schemes for split manufacturing insecure.

A number of techniques have emerged to defend against proximity attacks, including a pin-swapping-based countermeasure was proposed in [92]. Another approach reported in [93] indicates that if the layout is split at metal 1 layer, it will provide complete protection against proximity attacks; however, such technique is not feasible in practice as it will require state-of-the-art fabrication facilities to be available at the trusted BEOL foundry, which defeats the purpose of outsourcing IC manufacturing, hence impractical. More practical techniques proposed in [94] are based on randomization of the placement using graph colouring and clustering gates of the same type. Combining split manufacturing with logic locking is another interesting solution that enhances its security [94].

Despite reported attacks, split manufacturing remains an attractive solution, and it is predicted that it is going to play a vital role in the protection against IP piracy, once it starts to be deployed more widely.

#### Hardware Obfuscation

This approach consists of obscuring and hiding the structure and functionality of an IC design, which makes it very difficult for an adversary to steal the intellectual properties of a design, even if they have access to the netlist. Hardware obfuscation can be carried at different points of the IC design process as shown in Fig. 1.8. The following subsections discuss the techniques applied at each of these stages.

1. *Sequential logic locking*

   This form of obfuscation typically takes place at the pre-synthesis stage; it requires modifying the functional specification of the design. For sequential circuits defined as a finite state machine (FSM), a number of additional stages can be added, which are referred to as dark states. This renders the design non-functional, unless the correct key is supplied, an example of this approach was proposed in [95], although in this case the aim was to protect against Trojan insertion. Other forms of this approach include the dynamic state deflation approach in [96], the interlocking obfuscation in [97], and the active hardware metering discussed above. Recently the authors of [98] demonstrated that the above-mentioned techniques are not sufficiently secure, and they proposed an approach to obfuscate sequential circuits by replacing part of the design with a programmable logic circuit, which can protect the original design from accessing

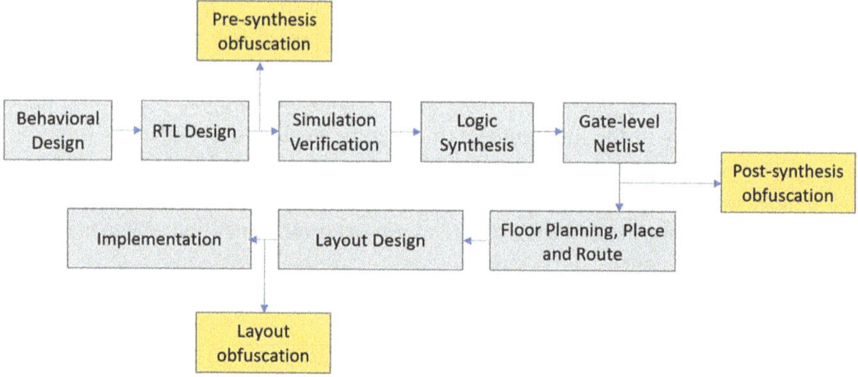


**Fig. 1.8** IC design flow with hardware obfuscation

by the foundries. Once the chips are manufactured and handed to a trusted testing facility, the logic is programmed with the intended functionality; a similar idea has been previously suggested in [99]. The use of configurable logic may be effective, but it is not very practical for a large volume chip production.

Another obfuscation technique that can be applied before synthesis was proposed in [100], wherein the RTL design is first transformed into a technology-independent gate-level description, and the functionality of the resulting gate-level netlist is then changed through modification of its state transition function, such that normal operation of the circuit can only be obtained after the successful application of a correct initialization sequence.

2. *Combinational logic locking*

This form of obfuscation typically takes place at the post-synthesis stage, it requires modification of the netlist file, for example, by randomly inserting multiplexers or XOR/XNOR gates (referred to as the key gates) in the combinational circuit, which render the design non-functional unless the correct logic values are applied to the inputs of these additional gates. A simplified example of this approach is shown in Fig. 1.9, wherein the insertion of the XOR gate will lead to an erroneous output, unless the correct key value is applied (in this case $k = 0$).

Early logic techniques have mainly focused on how to choose the location to insert the extra gates. Thus, several selection algorithms have been proposed:

(a) *Random logic locking*. This is where a location in the netlist is chosen at random and if there is not already a key gate in that location, then a key gate is inserted there [80]. Random insertion has many limitations since many of the key gates may be inserted far away from the critical path, and therefore do not cause much change to the output. Another limitation is that key gates are often isolated, making it easy for an attacker to determine the value of that key bit. Benefits of using this approach are the simplicity of its implementation and the speed of gate insertion.

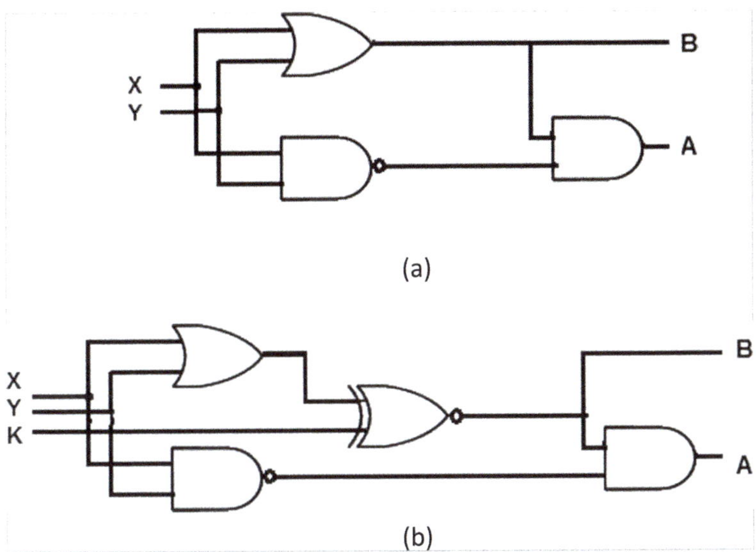


**Fig. 1.9** Combinational logic locking with an XOR gate: (**a**) original circuit and (**b**) locked circuit

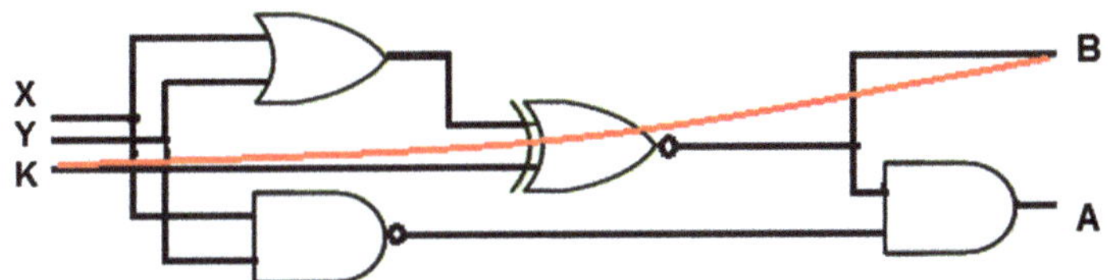


**Fig. 1.10** Circuit locked using XOR gate by applying the input pattern XY = "00", an attacker can sensitize key K to the output B

(b) *Fault analysis-based selection* wherein the errors caused by suppling the wrong inputs to the key gates can be assumed as stuck-at-faults [101]. Similar to the stuck-at-fault, not any wrong value caused by inserting the wrong key can corrupt the circuit output, as these can be masked by certain input patterns or other faults in the circuits. This type of locking will search for insertion locations that allow the error caused by wrong key insertion to propagate to the output, hence invalidating the functionality of the circuit. In this context, the fault impact (FI) metric is used, which evaluates the impact of every location in the circuit on error propagation, and the position with higher FI value will be selected for inserting the key gates.

(c) *Strong logic locking* algorithm was proposed to overcome a security vulnerability in prior locking approaches, which makes it feasible for an attacker to deduce individual bits in the key, as shown in Fig. 1.10.

Strong logic locking overcome this problem by inserting key gates in locations such that they form an interference pattern in the circuit, which makes it very difficult for an attacker to determine individual bit values of the

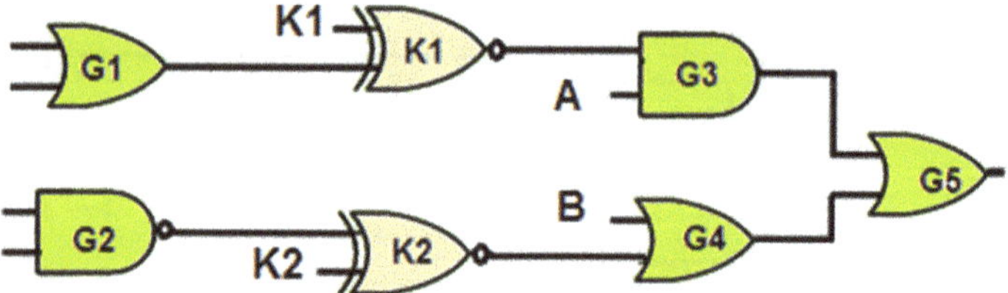


**Fig. 1.11** An example of concurrently mutable convergent key gates

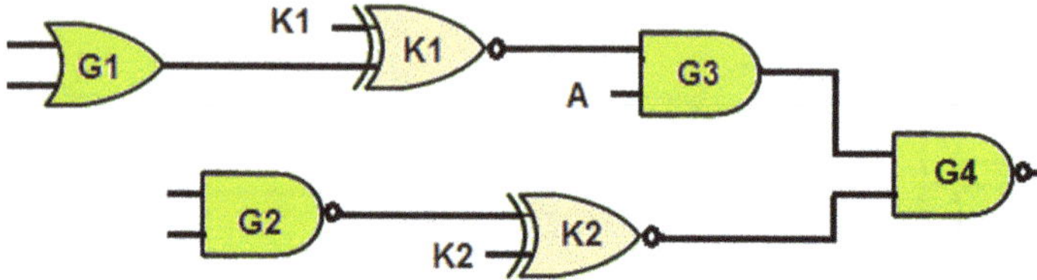


**Fig. 1.12** An example of sequentially mutable convergent key gates

key. This is achieved by ensuring that the key gates are inserted in locations that satisfy two main conditions. The first is that key gates must converge at some point. This is required to produce the interference pattern that makes this technique effective. The other condition is that the key gates must be "non-mutable", meaning that for each pair of key gates, neither of their key bits can be determined by applying an input pattern that "mutes" the effect of the other key gate. Gates that can be determined to be convergent (i.e. they satisfy the first condition) can be classified into three categories:

- Concurrently mutable convergent gates: an example of this is shown in Fig. 1.11, wherein the value of K2 can be worked out by muting the effect of K1 at gate G3, thus sensitizing K// to the output. Similarly, the value of K1 can be found by muting K2 at gate G4 and sensitizing the key bit to the output.
- Sequentially mutable convergent key gates: as shown in Fig. 1.12, wherein the value of K2 can be determined because K1 can be directly muted by an attacker. However, K2 cannot be directly muted, so the value of K1 has to be determined first before an attacker can determine a suitable input pattern to sensitize K2 to the output. If no such pattern exists, then the attacker can apply a brute force approach to just that set of key gates to determine the correct key bits.
- Non-mutable convergent key gates: If the neither of the key gates can be muted, then they are call non-mutable convergent key gates. As an attacker cannot access key inputs nor can they mute either gate in the pattern, they have to use a brute force attack. Therefore, it is preferred to use non-mutable convergent key gates wherever that is possible.

The main form of attack on the above discussed combinational locking technique rely on SAT-solving [102]. The attacker is assumed to have access to the locked netlist (e.g. Through reverse engineering a chip) and has obtained a working unlocked device (e.g. by purchasing from the market). STA-based attacks use distinguishing input patterns (DIPs) to iteratively eliminate incorrect keys, by comparing the output of the locked netlist and the working chip, respectively. This will make it feasible to obtain the set of equivalence class of correct keys, which contains all keys that do not corrupt the input–output relation of the encrypted circuit for all input vectors. To thwart this form of attacks, the SARlock approach was proposed in [103], it consists of introducing additional circuitry to inject masking signals if the inserted key is wrong, this maximizes the required number of distinguishing input patterns to recover the secret key, rendering the attack effort exponential to the number of key bits. Another defence against SAT attacks is the Anti-SAT technique [104], which also lead to an exponential increase in the number of required SAT attack iterations with the key size.

Another class of attacks that relies on netlist analysis was reported in [105]; it is based on the observation that the application of an incorrect key often results in a circuit with significantly more logic redundancy compared to the correct circuit, it was shown that more than half of the bits of the internal key can be found by identifying redundant logic. Most notably, this new attack can be performed solely on a locked netlist without an activated IC or a functional model. It is noteworthy that although there are now strong logic locking mechanisms such as those reported above, research in this area has become an arms race wherein novel attacks are emerging frequently, followed by corresponding countermeasures, hence it is safe to expect that this great level of research activities to continue.

3. *IC camouflaging*

This obfuscation technique performed at layout. It aims to protect against reverse engineering attacks that use image processing techniques [106, 107]. It works by introducing cells that appears identical from the top view but can implement one of many possible Boolean functions [108], which makes it infeasible for a reverse engineering attack to deduce the functionality of these disguised cells by analysing layout images. This type of obfuscation is performed by inserting additional structures in the layout such as filler cells, dummy contact or programmable standard cells.

This approach suffers from two major drawbacks. The first that the majority of existing camouflaging techniques is proven to be insecure against SAT-based de-camouflaging attacks as well as removal attacks [109–111] based on structural and functional information. Such attacks have sparked the emergence of more resilient camouflaging methods such as that reported in [112], which promises to overcome said security vulnerabilities. Another downside of IC camouflaging is its significant overheads, which has so far greatly limited its use in commercial products.

It is worth pointing out here that the cost associated with this approach should be balanced against the potential losses caused by loss of IP.

IP Watermarking

This the most widely deployed approach of mitigating methods against the risk of IP piracy. It consists of embedding the IP owner's signature into the design. This can be later used to prove the ownership of the design in the case of a legal dispute with a potential adversary [113].

#### 1.7.2.2 Countermeasures for Data Theft

Countermeasures for Side-Channel Analysis

In order for a SCA attack to succeed, the measurement of side-channel data must have sufficiently high signal-to-noise ratio (SNR), therefore countermeasures against this form of attacks aim to reduce the SNR as much as possible. This is achieved by reducing the signal level using leakage reduction methods or by increasing the level of the noise using noise injection techniques.

1. Leakage reduction approaches

   One of the fundamental causes of information leakage is the dependency of side-channel information on circuit's inputs, in conventional complementary metal–oxide–semiconductor (CMOS) implementations. Therefore, several countermeasures are aimed to mitigate the systematic properties of CMOS. Examples of these include the power balanced (*m*-of *n*) circuits [114, 115], the time enclosed logic (TEL) [116], current mode logic [117, 118] and randomized multitopology logic (RMTL) [119].

   A second technique to decrease the information leakage is the use of asynchronous design techniques, which removes the need for a global clock, therefore the computation operations become desynchronised making it harder to build correlation between the data being processed and measured side-channel information such as electromagnetic emission or power traces [120].

   A third approach is based on the use of voltage regulators, wherein existing on-chip power management schemes are tailored as a countermeasure against differential power analysis, this is achieved by randomizing the supply voltage and/or the frequency to break the one-to-one relationship between these parameters and the actual workload [121, 122]. A fifth approach reduces information leakage by decoupling the main power supply from the internal power supply that is used to drive logic gates, this is achieved with the help of buffering capacitances integrated into semiconductor [123].
2. Noise injection methods

Countermeasures based on noise injection can be applied at hardware level such as the work in [124], wherein the noise injection (NI) circuit consists of a two-level stochastic linear feedback shift register (LFSR) stage followed by a 5-bit current steering Digital to Analogue Converter (DAC). Another hardware-based approach is to randomize the clock frequency [125].

Software-based countermeasures include random insertion of dummy instructions into the execution sequence of algorithms [126, 127], shuffling, random pre-charging [128, 129], and Boolean and arithmetic masking [130].

It is worth noting here that the use of leakage reduction/noise injection techniques on its own cannot prevent side-channel attack, as adversaries may resort to using advanced signal procession techniques that will help the SCA attack to be successful even with small SNR. However, these approaches can still make SCA attack harder, and in some case prohibitively expensive.

3. Other techniques

One effective approach to protect against side-channel information is to key update. The latter can prevent the adversary from accumulating sufficient amount of side-channel information to guess the key; this is achieved by frequently changing the key [131]; in some cases, the key is only used once in each cryptographic operation [132].

In practice, a combination of the above countermeasure may be needed to increase the resiliency of the system against side-channel analysis.

#### Countermeasure for Cache Timing Attacks

The most intuitive approach in this case is to hide timing information [33], which can be achieved by injecting random delays into the execution of a task, which makes it harder to deduce the correlation between the secret data and the memory access time or by fixing the computation times of all operations, which will help reduce the resolution of the measured timing measurements and foil the attack; however this comes at relatively high cost in terms of performance or computing resources. One example of constant time techniques is to adopt a data oblivious memory access technique, which breaks the correlation between the data being read and the memory access patterns. A simplistic implementation of this technique, in the case of AES system, is to read all the entries from the lookup table and only use what is required. Practical implementation of this approach includes the work in [133] that developed a programme transformation for hiding memory accesses at the cost of an increase in the memory and time overheads. An example of noise injection techniques has been shown in [134], which relied on the use of a variety of techniques including the randomization, compacting and preloading of AES lookup tables, and this led the distribution of AES execution times to have a Gaussian distribution over random plaintexts; however this caused more than 50% degradation in performance. Other types of countermeasures rely on explicit control of cache access by different processes, for example by flushing all level of caches during virtual machines switches in cloud computing when the

CPU switches between different security domains [135]. Other techniques rely on disabling memory page sharing [136], or disabling hyperthreading [137]. The latter has become a common practice on public cloud services, such as Microsoft's Azure. Cache partitioning is another method used to protect against timing attacks. There are a number of implementations of this approach, including partitioning the L1 cache between threads to eliminate the cache contention [137]. Another example of this approach is the use of Cache colouring techniques to allow portioning the cache at the software level [138].

Despite the availability of a wide range of mitigations techniques, latest studies indicated that cache timing attacks still present a threat to internet of things devices either due to the lack of appropriate protection mechanisms and the residual information leakage that still persists [139].

Given the relative simplicity of waging this attack, even remotely [140], and the significant proliferation of internet of things devices, some of which controlling critical infrastructure, it is extremely likely this will be one of the main techniques used by adversaries to compromise the security of computing devices.

#### Countermeasures for Speculative Execution Attacks

This type of attacks has violated fundamental assumptions underpinning numerous software security mechanisms, including operating system process separation, and containerization, just-in-time (JIT) compilation [29]. Therefore, its implications on existing systems and long-term impact on the design of computer architectures are still being analysed, with new attacks vectors emerging continuously. Equally, defence mechanisms have started to emerge shortly after Spectre, the first known attack of this type, was announced. Intel has initially proposed three types of mitigation techniques [141]:

(a) *Bounds check bypass mitigation*, which consists of modifying the software to insert barriers in order to stop speculation in appropriate places, in particular, an LFENCE instruction is recommended. The locations where such an instruction is to be inserted will be software-dependent, Intel analysis of Linux kernel for example has only found few of places where LFENCE insertion is required.
(b) *Branch target injection mitigation*, wherein two techniques have been devised. Only one of which needs to be used, which can be decided based on the requirements of the system under consideration. The first is based on the insertion of an interface between the processor and system software, which provides mechanisms that allow system software to prevent an attacker from controlling the victim's indirect branch predictions, such as flushing the indirect branch predictors at an appropriate time. The second technique relies on modifying the software to replace indirect near jump and call instructions with an alternative code sequence. The latter pushes the target of the branch in question onto the stack and then executes a Return (RET) instruction to jump to that location; this approach is referred to a "return trampoline", aka "retpoline".

(c) *Rogue data cache load mitigation*, wherein the operating system software is modified to ensure that privileged pages are not mapped when executing user code in order to protect against user mode access to privileged pages.

More recently researcher from Intel have proposed another countermeasure [142], which is based on the use of a new type of memory called "Speculative Access Protected Memory".

When the central processing unit (CPU) detects a memory access targeting this protected RAM, it starts processing instructions serially and refuses to engage in speculative execution *until* the SAPM-targeting instruction has been retired. Applications can still store secret data in SAPM as required, but this will be at the cost of an increase in the latency of memory access operations.

In principles, the applications of any of the above technique will lead to a performance degradation; this is deemed a necessary price of having additional security protections.

Additionally, a number of countermeasures have also been proposed by academia. In [143], the authors developed a data centric mitigation mechanism for the bounds check bypass variant of Spectre, it is based on the observation that identifying a programme's secret data can be easier than identifying vulnerable code blocks (i.e. Spectre gadgets), as the latter which can appear in any branch of the programme even if it does not process sensitive information. Consequently, this countermeasure consists of identifying the areas in the memory that contain secret data, and marking these as *non-speculative memory regions by adding a* micro-architecture extension and corresponding software interface to the operation system. In [144], another data-based approach is proposed, wherein the speculative data obtained from the first phase of the attack is shielded from any potential leakage. This is achieved by blocking the use of such data by all dependent instructions until the source instruction is determined to be safe, such an approach will lead to approximately 18% performance degradation. The work in [145] proposes the concept of conditional speculation wherein instructions which are deemed unsafe are not allowed to execute speculatively. The work in [146] proposes a defence mechanism that isolates caches' hits, misses, and metadata updates across protection domains, making it harder to perform cache timing attacks, an integral part of speculative execution attacks, which makes the latter unlikely to succeed. In [147], a hardware-based detection mechanism for Spectre and Meltdown is developed, it uses data from the hardware performance counters such as cache misses and branch miss-predictions.

Finally, it is important to remember that the speculative execution attacks rely on a malware running locally. The latter is assumed to have been installed by an adversary that has a physical or a remote access to the machine. As such, the simplest countermeasure is to stop the installation of the malware in the first place; this can be achieved by keeping software up-to-date and avoiding suspicious downloads.

(d) Countermeasures for PUF modelling attacks

Techniques to prevent modelling attacks are either circuit-based or protocol-based. The latter relies on obfuscating the relationship between the challenges and the responses at the protocol level. This approach has a fundamental flaw as it cannot protect against invasive physical attacks, such as the reliability-based Covariance Matrix Adaption (CMA)-ES attack [148], which is capable of breaking even highly output obfuscated PUF protocols, e.g. Slender PUF. Circuit design techniques include two types of solutions. The first consists of adding extra blocks at the output/input of the PUF (e.g. Hash function or other cryptographic primitives [149]). The main weakness of this approach is its limited capability to protect against invasive physical attacks; this is because the PUFs used in these solutions to date are not resilient to modelling attacks, so ultimately if an adversary manages to access the input/output of the PUF, the security of such a system will break. A more promising approach relies on increasing the complexity of the PUF circuit itself. There are a number of ways this can be achieved. In [150] the use of sub-threshold voltage design was shown to improve the resilience against modelling attacks, but it did not completely prevent it. Other techniques include combining the output of two PUFs such as the XOR PUF; however, this design has already been broken [72]. Recent work has indicated that combining multiple PUF primitives leads to stronger resilience to machine learning modelling attacks. Examples include the multiple PUF approach in [151], a cascaded architecture [152] and the interpose PUF [153].

## *1.7.3 Countermeasures for Sabotage Attacks*

### 1.7.3.1 Countermeasures for Cyber-Physical Attacks

Defence mechanisms used to protect from regular malware attacks are useful in this case. These include proper management and regular updates of firewalls; packet-sniffer software; virus checkers; access control lists [154]. Other mitigation techniques that are particularly relevant to the Stuxnet attack is robust virus scanning of portable media devices (e.g. USB) or banning these all together, the use of endpoint security software to prevent the propagation of malware through over the network, and the use of firewalls to separate industrial from general business network. More advanced protection techniques also include the use of machine learning algorithms to monitor network activities and detect anomalous traffic [155].

In addition to the above, countermeasures should address the security threat of human users, who might intentionally or unintentionally breach the security of the system by downloading malware. Therefore strong user authentication approaches should be adopted such as the biometric-based user identification method proposed in [156]. It is worth noting here that the development of an effective cyber-physical attack of the scale and sophistication of Stuxnet requires a great deal of resources, which is only available to state-sponsored organizations, which makes it less likely compared to other forms of cyber-physical attacks.

#### 1.7.3.2 Countermeasures for Rowhammers Attack

There are a number of defence mechanisms that can be implemented in this case. The most intuitive approach will be to use DARM technologies with less leakage currents, in practice, this may be prohibitively expensive or not even feasible. Another approach would be to control repeated access to the same memory raw by limiting the number of accesses and access intervals, again this solution may incur a large overhead in terms of cost, power, performance and/or complexity. Other techniques include the use of error correction techniques; however, such an approach has been shown to still have security vulnerabilities [157]. A more practical approach is to increase the memory refresh rate, hence reducing the likelihood of row hammer-induced bit flips. This technique, which can be implemented at the software level, has already been adopted in commercial electronics such as Apple's products.[5]

#### 1.7.3.3 Countermeasures for CLKscrew Attacks

There are a number of defence mechanisms that can be implemented in this case. The first approach is enforcing strict operation limits on the hardware regulators using e-fuses or dedicated circuitry [47], although effective, this approach may be hard to implement in practice as hardware regulators are designed to work across a wide range of processor, for example, the same regulator is used for PMIC and Galaxy Note 4. Another approach is to use fault tolerance techniques to mitigate the effects of erroneous computations due to induced faults. These can be implemented using information redundancy (e.g. error correction codes) or hardware or time redundancy combined with majority voting circuitry. More practical countermeasures can be deployed at the software level [158], these include the compilation of sensitive code multiple times and the use of checksum integrity verification, these techniques are already implemented in highly dependable systems; however, its use in resource-constrained devices has so far been limited due to its impact on energy dissipation.

---

[5]https://support.apple.com/en-gb/HT204934.

### 1.7.4 *Countermeasures for Tampering Attacks*

#### 1.7.4.1 Countermeasures for Hardware Trojan Insertion

Protection against hardware Trojans can be approached by different techniques. A brief description of these techniques and some examples are presented in the following.

Trojan Prevention Techniques

The goal of these techniques is to modify the original circuit to either assist another countermeasure technique or obfuscate the original logic to make harder the insertion of a Trojan. One way of inserting Trojans at the layout level is by replacing the filler cells of the IC by cells that perform a malicious operation. To avoid this, filler cells can be replaced by functional cells to implement an LFSR/MISR-like circuit that generates a digital signature of the IC such that if any of its cells are modified because of a Trojan insertion, the digital signature changes [159].

Another technique uses the empty area of the IC (filler cells) to replicate logic that processes sensitive data of the system. The replicas are spread over the IC, and only one is activated randomly or through a keyed physically unclonable function (PUF) after the device is received from the foundry [160]. If a Trojan is inserted in only one of the replicas, it is unlikely that it can perform the malicious operation because that replica may not be activated. One obfuscation technique uses Linear Complementary Dual (LCD) codes to encode and masks all the internal registers of the original circuit such that the modification of less than a certain number of registers of the encoded circuit does not have an effect on the original circuit [161]. A similar technique uses Linear Complementary Pair (LCP) of codes to encode the original circuit using two parameters such that if a Trojan modifies less than $d_{\text{trigger}}$ registers of the encoded circuit, it does not disclose valid information of the original circuit, and if the trojan modifies less than $d_{\text{Payload}}$, it would generate wrong codes [162]. These techniques also protect the original circuit against probing, fault injection and side-channel attacks.

Trojan Detection

Detection is the most common Trojan countermeasure found in the literature. The general idea is to determine at some stage of the lifecycle of a system if a Trojan has been inserted in it.

1. Pre-silicon detection

   Pre-silicon techniques are used to detect Trojans in the netlist or RTL of a design before manufacturing. These techniques can be applied when IP from an untrusted vendor is used in a new design; however, they are limited by the availability of the HDL code of the IP since in some cases the IPs are encrypted. Some of these techniques use structural analysis to detect Trojans in the netlist or RTL by identifying rarely used circuits and input combinations during verification.

UCI (Unused Circuit Identification) technique [163] focuses on identifying parts of the circuit that do not affect outputs during testing. It uses data flow graphs to check if a dependent signal does not change on variations of a driving signal during simulation. The circuit in between is considered unused and therefore potentially malicious.

FANCI targets small well-hidden trigger circuits of Trojans which could evade testing and code inspection [164]. It works by analysing the truth table of all gates in order to find inputs that have low impact on the output signals. A metric is used to measure the output signal activity of the gate and if its value is lower than a user-defined threshold, the gate is considered malicious. One characteristic of this technique is that the less well-hidden the trigger is, the less likely it is to detect it.

VeriTrust identifies unused input combinations of flip-flops and primary outputs which can be potentially related to a Trojan trigger circuit [165]. This technique has some limitations; for example, Trojans that are triggered by normally used combinations of the inputs might not be detected, this could be the case of Trojans that induce a malfunction in the circuit. Another limitation is that only functionally activated Trojans can be detected, in other words, always-on Trojans are not detected. Formal verification techniques are used to detect Trojans by ensuring that the IP implementation represents exactly the IP specification.

Formal methods can evaluate if the system can be operated in unspecified ways that are usually exploited by Trojans (illegal input data, forbidden system states, etc.) [166]. Formal methods can also be used to check security properties that evaluate if the system leaks sensitive information (cryptographic key, plaintext or intermediate computation) [167]. One limitation of these techniques is that they can check if a design satisfies pre-defined security properties, but not if the design has additional hidden vulnerabilities. Another way of Trojan detection in pre-silicon stages is by inspecting the RTL code or by reverse engineering the netlist of a system in order to identify suspicious operations in the implementation of the system. Although this is a manual and time-consuming technique, it could be very effective to detect hardware Trojans.

2. Post-silicon (test-time) detection

   Most of the detection techniques are designed to detect Trojans during testing of the integrated circuit once it is manufactured (post-silicon detection). Many of these techniques rely on the existence of a golden IC (Trojan-free IC) for comparison in order to identify Trojan-infected ICs.

   There are two main categories of methods for Trojan detection during testing: invasive methods and non-invasive methods.

   In an invasive or destructive method, an IC is de-packaged to take images of its layout layers and compare them against the golden IC. Although this technique can be very effective detecting Trojans, it is very expensive and complex. Additionally, after an IC is de-packaged and delayered, it cannot be used again, and, even if a Trojan is not detected on a particular IC, it is not possible to guarantee that other ICs of the batch are not infected.

Some invasive methods do not require a golden IC, for example, a technique compares simulated images against real images of the IC obtained using near-IR wavelengths to detect Trojans inserted in filler cells by noticing that filler cells are more reflective than functional cells [168].

Non-invasive or non-destructive methods use characteristics of the IC that can be measured during testing to detect Trojans by observing deviations from their expected behaviour. Trojans can be detected using functional and structural testing; however, since Trojans are normally activated under rare conditions, test patterns used for detecting defects on an IC are not effective for Trojan detection. Some techniques have been developed to generate test patterns that facilitate the activation of such rare conditions in order to observe the effect of the Trojan in primary outputs [169, 170]. Although these techniques can be effective in some cases, some Trojans do not change the functionality of the circuit, instead, they change non-functional specifications or insert antennas to leak data that cannot be detected by functional or structural testing. Other non-invasive techniques are based on side-channel analysis. The basic idea is to measure the side effects that Trojans have on certain parametric characteristics of the IC, for example, path delay [171], power consumption (current integration [27, 172] and transient currents [173, 174]), electromagnetic radiation [175] and temperature [176, 177]. One limitation of most of the methods that use these techniques is that they require a golden IC as reference. In general, one challenge of the detection methods based on side-channel analysis is that the side effects of low-overhead Trojans can be masked by the effects of process and environmental variations in the IC. Another challenge is to design measurement techniques that can cover all the area and every network of the chip to detect Trojans inserted in any part of the IC.

3. Runtime detection

Some attacks can exploit a combination of hardware and software Trojans to perform malicious operations. In this case, it might be possible that test-time and pre-silicon detection techniques fail at finding hardware Trojans because they do not perform directly a malicious operation, and therefore can be hiding more easily. Hardware property checking (HPC) modules can be used to detect such Trojans by checking high-level and critical behavioural invariants at runtime. For example, in a processor, the HPC can check the integrity of on-chip buses rules and the correct management of supervisor/user modes [178].

One disadvantage of this technique is the overhead, especially in lightweight applications where area and power are constrained. Runtime Trojan detectors are also vulnerable to Trojan insertion or modification at layout level. In [178] the authors present solutions to this problem: designing the IC using 3D technology such that the target circuit and the Trojan detector are fabricated in different dies by different foundries; implementing the Trojan detector using reconfigurable logic (FPGA); or using encoded circuits to protect the Trojan detector as described in [162]. Other techniques implement security policy checkers that enable runtime monitoring of security issues presented in untrusted IPs that can lead to system-level Trojan threats in SoC [179].

4. Trojan runtime protection

   Other countermeasures techniques are designed to circumvent the malicious operations of Trojans at runtime without the need to detect or remove them. For example, hardware wrappers can be used to enforce security policies of interest in untrusted IP modules within SoCs [180]. Other mechanisms can be used to thwart Trojans that use side-channel effects to leak information by randomizing the data bus of SoC such that attacker cannot find any correlation between what he measures and what the Trojan transmits [181].

#### 1.7.4.2 Countermeasures for Fault Injection Attacks

The majority of fault injection techniques can be implemented with relatively cheap equipment and tools, WHICH makes it harder to completely prevent this type of attacks. Nevertheless, there are a number of effective mitigation approaches, which can be classified in the following categories.

(a) Tamper-resistant techniques

   These types of approaches limit physical access to the device, hence making fault injection more difficult. These include the use of hardened steel enclosures, locks, encapsulation or security screws.

(b) Tamper detection

   The use of physical encapsulation is sometimes combined with tamper detection mechanisms to provide better protection. This can be achieved using sensors, switches or dedicated circuitry. Sensors are used to detect all the above-mentioned types of fault injections attacks; these include changes in the environment temperature, level of radiation and fluctuations of power and voltage supply. Switches are used to detect a breach of the physical boundaries of the device such as opening the enclosure. Examples of these are pressure contacts, magnetic and mercury switches. More expensive temper detection techniques consist of using dedicated circuitry that are wrapped around the device. Examples of this include the use of nichrome wire, whose resistance can change if its dimensions are physically modified. Another example is the use of fibre optic cables that alternate the light power travelling through them if tampered with [13]. Similar approaches include the use of an active mesh of wires that is driven by a random number generator, wherein physical tampering is detected by monitoring the plurality of the number generated [182]. The use of wire mesh has also been proposed in [183], wherein conductive mesh is affixed onto interior surfaces of the outer housing of the device, which are combined with a detection mechanism within the device. This approach allows the detection of open or short-circuit condition that results from an unauthorized opening of the device enclosure. Physically unclonable functions (PUF) have also been suggested [184] by monitoring the changes in PUF responses resulting from physical modification of its structure.

(c) Tamper response mechanisms

The detection of a tampering attempt will typically lead to a response from the underlying system to stop the intrusion. There are a variety of responses in this case; these include deletion of sensitive information from the memory, shutting down of the device or complete physical destruction [185]. Self-destruct electronics technologies are now available commercially, for example the British company Secure Devices have developed a solid-state hard drive that can self-destruct [186].

(d) Fault tolerance design

This technique aims to modify the design of the cryptographic algorithms such that injected faults can be detected and potentially corrected [187]. This can be achieved using information redundancy (e.g. error correction codes) or hardware or time redundancy combined with majority voting circuitry.

Software-based defence mechanisms may also be relevant in this case to mitigate against computation errors induced by Clkscrew. These include execution redundancy (executing sensitive code multiple times) and compiling code with checksum integrity verification [158].

## 1.8 Security Validation

The purpose of this step is to establish whether or not the implemented countermeasures have successfully pacified the threats identified in step 2. This require a careful examination of the attack scenario, capabilities of the adversary and the protection level provided by the countermeasure in question. To achieve this, one needs to systematically consider the following questions for each identified attack and corresponding countermeasures:

1. Who are the most likely adversaries?
2. Has the countermeasure removed the root of vulnerability that made this attack technically feasible?
3. Has the countermeasure increased the difficulty of the attack beyond the resources available to the most likely adversary?

In order to illustrate how this process can be used, we will consider the following examples:

### 1.8.1 Case Study 1: The Use of Obfuscation to Prevent PUF Modelling Attacks

Envisage a wireless sensor network used for the monitoring and control of other electronic appliances, wherein each sensor incorporates a PUF as its identity generator such that the challenges and responses of the PUF are used for device authentication.

The goal of the adversary is to develop a software model of the PUF in one of more network devices, to forge identities and secure unauthorized access to resources. To achieve this, they need to collect a sufficient number of challenge/response pairs as discussed previously.

To mitigate the risk of this attack, the designer implements a challenge obfuscation technique using additional cryptographic circuitry such as hash functions.

To validate the security of this obfuscation approach, this analysis considers three application scenarios, a smart home, an industrial site and a governmental institution (e.g. an embassy). For each case the three main questions listed above are considered.

1. Who are the most likely adversaries?

   For the smart home scenario, this is most likely to be a class one, i.e. a small criminal group with relatively limited resources. On the other hand, attacks on industrial site are most probably waged by larger criminal groups (i.e. class three), whereas government buildings are mostly targeted by well-funded institutions (i.e. class four).
2. Has the countermeasure removed the root of vulnerability that made this attack technically feasible?

   The root of vulnerability in this case is the correlation between the response and the challenge of the PUF, which makes it learnable by machine learning algorithms.

   The use of obfuscation method does not remove this vulnerability. In fact this attack is still technically feasible because, for example, if an adversary can secure access to the internal wiring of the device in order to apply stimulus to the inputs and the PUF and measure its output, they can still create a software clone that mimics its behaviour to a high degree of accuracy.
3. Has the countermeasure increased the difficulty of the attack beyond the resources available to the most likely adversary?

   Providing an answer to this question requires knowledge on the difficulty of waging the attack, its practical and economic viability after implementing the countermeasures. In the case under consideration, the attack is still technically feasible as argued above, but requires specialized tools (e.g. chip delayering and internal probing), therefore the difficulty is classified as (3).

   The viability of the attack depends on the available resources of the most likely adversary and the potential benefits that can be gained from waging the attacks.

**Table 1.10** Security validation for using logic obfuscation to Thwart PUF modelling attacks

| Application scenario | Most likely adversary | Root of vulnerability removed | Difficulty of the attacks after implementing the countermeasure | Is the attack still viable? |
|---|---|---|---|---|
| Smart home | Class 1 | No | 3 | No |
| Industrial site | Class 3 | No | 3 | Yes |
| Governmental institution | Class 4 | No | 3 | Yes |

In the case of a smart home scenario, the use of obfuscation techniques will increase the required resources beyond the reach of a type 1 adversary, which makes it non-viable.

For the industrial site or government building scenarios, the potential benefits (economic or otherwise) make it worthwhile investing more resources to wage a more invasive PUF modelling attack. A summary of the analysis is provided in Table 1.10.

### *1.8.2 Case Study 2: The Use of Logic Locking to Prevent IP Theft at the Manufacturing Stage*

The outsourcing of IC manufacturing, driven by cost saving measures, has increased the risk of IP piracy during the fabrication stage.

The goal of the adversary in this case is to expose design secrets by reverse engineering the chip and extracting the netlist. To mitigate the risks of such an attack, the designer implements one of the logic locking mechanisms to obfuscate the functionality of the design. It is typically assumed that the unlocking key is stored securely on chip after unlocking of the IC by a legitimate IP owner; in some cases, every single chip has its own unlocking key, which can be achieved using PUF technology or other forms of hardware-based identity generators.

To validate the security of this approach, this analysis considers two generic scenarios consumer electronics and military-scale devices. For each case, the three main questions listed above are considered.

1. Who are the most likely adversaries?

   For the consumer electronics case, this is most likely to be a class three, i.e. a large criminal such as a competitor. On the other hand, attacks military-scale devices are most probably waged by state-sponsored actors (i.e. class four).
2. Has the countermeasure removed the root of vulnerability that made this attack technically feasible?

   The root of vulnerability in this case is the correlation between circuit layout and the gate-level netlist and ultimately the design functionality. Although the use of logic locking mechanisms gives the IP holder more control over the design,

**Table 1.11** Security validation for using logic locking to Thwart IP theft

| Application scenario | Most likely adversary | Root of vulnerability removed | Difficulty of the attacks after implementing the countermeasure | Is the attack still viable? |
|---|---|---|---|---|
| Consumer Electronics | 3 | No | 3 | No |
| Military hardware | 4 | No | 3 | Yes |

they do not remove this vulnerability, In fact, IP piracy of a locked circuit would still be technically feasible if an attacker has sufficient resources as argued [188]; these include the gate-level netlist of the locked design, multiple locked ICs, multiple unlocked ICs, lithographic masks and other fabrication artefacts used to manufacture the ICs, and the state-of-the-art IC analysis equipment.

3. Has the countermeasure increased the difficulty of the attack beyond the resources available to the most likely adversary?

   Providing an answer to this question require knowledge on the difficulty of waging the attack, its practical and economic viability after implementing the countermeasures. In the case under consideration, the attack is still technically feasible as argued above, but requires specialized tools (e.g. chip delayering and internal probing), therefore the difficulty is classified as (>5).

   The viability of the attack depends on the available resources of the most likely adversary and the potential benefits that can be gained from waging the attacks.

   In the case of consumer electronics, the most likely adversary is a class three attacker, with the financial capabilities that allows recruiting a number of disgruntled employees within an IC fabrication facility. This can give such an adversary access to the design's netlist and, multiple locked and unlocked ICs. These resources are not sufficient to break the state-of-the-art logic locking approaches available today.

   For military hardware, the potential benefits of IP piracy typically justify investing more resources to take a more invasive approach in order to reveal design secrets.

   A summary of the analysis is provided in Table 1.11.

## 1.9 Conclusions

This paper has presented a threat modelling approach specifically tailored for hardware supply chain-related risks, counterfeiting, information leakage, sabotage and tampering (CIST). A comprehensive review of attack mechanisms associated with each of these risks have been presented, including a classification of the difficulty of each attack, the capability of the assumed adversary, the root of vulnerability that made the attack feasible (technically or otherwise) and the consequences of the attack. In addition, the state-of-the-art defence mechanisms associated with each attack scenario have been discussed in details. Finally a security validation framework has been outlined, which allows evaluating the effectiveness of countermeasures based on the level of protection it can provide and the capabilities of the most likely adversary.

**Acknowledgment** The author would like to thank the Royal Academy of Engineering for supporting this research (Grant No. IF\192055).

## References

1. E. Council, Electronic Systems A Plan for Growth: One Year On, 2014
2. B.T. Horvath, All Parts are Created Equal: The Impact of Counterfeit Parts in the Air Force Supply Chain, Air War College, Air University Maxwell AFB United States, 2017
3. F-Secure, Attack Landscape 2019, 2020
4. S. I. A. a. S. R. Corporation, Semiconductor Research Opportunities An Industry Vision and Guide, 2017
5. J. Di, S. Smith, A hardware threat modeling concept for trustable integrated circuits, in *2007 IEEE Region 5 Technical Conference*, 2007, pp. 354–357
6. M. Rostami, F. Koushanfar, R. Karri, A primer on hardware security: models, methods, and metrics. Proc. IEEE **102**, 1283–1295 (2014)
7. Microsoft. *STRIDE*, 2007. Available: https://www.microsoft.com/security/blog/2007/09/11/stride-chart/
8. A. Varghese, A.K. Bose, Threat modelling of industrial controllers: a firmware security perspective, in *2014 International Conference on Anti-Counterfeiting, Security and Identification (ASID)*, 2014, pp. 1–4
9. T. Simonite, To keep pace with Moore's law, chipmakers turn to 'Chiplets'. *Wired*, 2019. Available: https://www.wired.com/story/keep-pace-moores-law-chipmakers-turn-chiplets/
10. R.J.E. Andrew, P. Moore, R.C. Linger, *Attack Modeling for Information Security and Survivability* (Carnegie Mellon University, Pittsburgh, PA, 2001)
11. B.L.P. Saitta, M. Eddington, Trike v.1 Methodology Document, Jul 2005
12. S.H. Weingart, S.R. White, W.C. Arnold, G.P. Double, An evaluation system for the physical security of computing systems, in *1990 Proceedings of the 6th Annual Computer Security Applications Conference*, 1990, pp. 232–243
13. J. Grand, Practical secure hardware design for embedded systems, in *Proceedings of the 2004 Embedded Systems Conference*, San Francisco, CA, 2004
14. F.L. Nawfal Fadhel, L. Aniello, A. Margheri, V. Sassone, Towards a semantic modelling for threat analysis of IoT applications: a case study on transactive energy, 2019
15. P. Missier, K. Belhajjame, J. Cheney, The W3C PROV family of specifications for modelling provenance metadata. Presented at the Proceedings of the 16th International Conference on Extending Database Technology, Genoa, Italy, 2013
16. World Wide Web Consortium, Prov-n: The Provenance Notation W3C Recommendation, 2013
17. M. Yampolskiy, P. Horvath, X.D. Koutsoukos, Y. Xue, J. Sztipanovits, Taxonomy for description of cross-domain attacks on CPS. Presented at the Proceedings of the 2nd ACM International Conference on High Confidence Networked Systems, Philadelphia, PA, 2013
18. *Apple's List of Suppliers* (2019, Mar 2020). Available: https://www.apple.com/supplier-responsibility/pdf/Apple-Supplier-List.pdf

19. B. Halak, *Ageing of Integrated Circuits Causes, Effects and Mitigation Techniques* (Springer, Cham, 2020)
20. U. Guin, K. Huang, D. DiMase, J.M. Carulli, M. Tehranipoor, Y. Makris, Counterfeit integrated circuits: a rising threat in the global semiconductor supply chain. Proc. IEEE **102**, 1207–1228 (2014)
21. E. Andrijcic, B. Horowitz, A macro-economic framework for evaluation of cyber security risks related to protection of intellectual property. Risk Anal. **26**, 907–923 (2006)
22. R. Quijada, A. Raventós, F. Tarrés, R. Durà, S. Hidalgo, The use of digital image processing for IC reverse engineering, in *2014 IEEE 11th International Multi-Conference on Systems, Signals & Devices (SSD14)*, 2014, pp. 1–4
23. G. Kim, M. Ma, I. Park, A fast and flexible software for IC reverse engineering, in *2018 International Conference on Electronics, Information, and Communication (ICEIC)*, 2018, pp. 1–4
24. D. Cheng, Y. Shi, B.-H. Gwee, K.-A. Toh, T. Lin, A hierarchical multiclassifier system for automated analysis of delayered IC images. IEEE Intell. Syst. **34**, 36–43 (2018)
25. D. Cheng, Y. Shi, T. Lin, B.-H. Gwee, K.-A. Toh, Hybrid {K}-means clustering and support vector machine method for via and metal line detections in delayered IC images. IEEE Trans. Circuits Syst. II Exp. Briefs **65**, 1849–1853 (2018)
26. J.J. Paul Kocher, B. Jun, Differential power analysis, in *CHES*, 1999
27. A. Moradi, O. Mischke, C. Paar, One attack to rule them all: collision timing attack versus 42 AES ASIC cores. IEEE Trans. Comput. **62**, 1786–1798 (2013)
28. Y. Hayashi, N. Homma, T. Mizuki, T. Aoki, H. Sone, L. Sauvage, et al., Analysis of electromagnetic information leakage from cryptographic devices with different physical structures. IEEE Trans. Electromagn. Compat. **55**, 571–580 (2013)
29. P. Kocher, J. Horn, A. Fogh, D. Genkin, D. Gruss, W. Haas, *et al.*, Spectre attacks: exploiting speculative execution, in *2019 IEEE Symposium on Security and Privacy (SP)*, 2019, pp. 1–19
30. M. Lipp, M. Schwarz, D. Gruss, T. Prescher, W. Haas, A. Fogh, et al., Meltdown: reading kernel memory from user space. Presented at the Proceedings of the 27th USENIX Conference on Security Symposium, Baltimore, MD, 2018
31. J.V. Bulck, M. Minkin, O. Weisse, D. Genkin, B. Kasikci, F. Piessens, et al., Breaking virtual memory protection and the SGX ecosystem with foreshadow. IEEE Micro **39**, 66–74 (2019)
32. A. Johnson, R. Davies, Speculative execution attack methodologies (SEAM): an overview and component modelling of spectre, meltdown and foreshadow attack methods, in *2019 7th International Symposium on Digital Forensics and Security (ISDFS)*, 2019, pp. 1–6
33. D.A. Osvik, A. Shamir, E. Tromer, *Cache Attacks and Countermeasures: The Case of AES* (Heidelberg, Berlin, 2006), pp. 1–20
34. D. Gullasch, E. Bangerter, S. Krenn, Cache games—bringing access-based cache attacks on AES to practice, in *2011 IEEE Symposium on Security and Privacy*, 2011, pp. 490–505
35. P. Stewin, I. Bystrov, *Understanding DMA Malware* (Heidelberg, Berlin, 2013), pp. 21–41
36. S. Skorobogatov, Data remanence in flash memory devices, in *International Workshop on Cryptographic Hardware and Embedded Systems*, 2005, pp. 339–353
37. Y. Kai, Z. Xuecheng, Y. Guoyi, W. Weixu, Security strategy of powered-off SRAM for resisting physical attack to data remanence. J. Semicond. **30**, 095010 (2009)
38. M.S. Mispan, M. Zwolinski, B. Halak, Ageing mitigation techniques for SRAM memories, in *Ageing of Integrated Circuits: Causes, Effects and Mitigation Techniques*, ed. by B. Halak, (Springer International Publishing, Cham, 2020), pp. 91–111
39. B. Halak, Hardware-based security applications of physically unclonable functions, in *Physically Unclonable Functions: From Basic Design Principles to Advanced Hardware Security Applications*, ed. by B. Halak, (Springer International Publishing, Cham, 2018), pp. 183–227
40. B. Halak, Security attacks on physically unclonable functions and possible countermeasures, in *Physically Unclonable Functions: From Basic Design Principles to Advanced Hardware Security Applications*, (Springer International Publishing, Cham, 2018), pp. 131–182

41. E.D. Knapp, J.T. Langill, Chapter 3—Industrial cyber security history and trends, in *Industrial Network Security*, ed. by E. D. Knapp, J. T. Langill, 2nd edn., (Syngress, Boston, 2015), pp. 41–57
42. K. Hemsley, R. Fisher, A history of cyber incidents and threats involving industrial control systems, in *Critical Infrastructure Protection XII*, Cham, 2018, pp. 215–242
43. Y. Kim, R. Daly, J. Kim, C. Fallin, J.H. Lee, D. Lee, et al., Flipping bits in memory without accessing them: an experimental study of DRAM disturbance errors. ACM SIGARCH Comput. Architect. News **42**, 361–372 (2014)
44. A. Tatar, R.K. Konoth, E. Athanasopoulos, C. Giuffrida, H. Bos, K. Razavi, Throwhammer: rowhammer attacks over the network and defenses. Presented at the Proceedings of the 2018 USENIX Conference on Usenix Annual Technical Conference, Boston, MA, 2018
45. M. Seaborn, T. Dullien, Exploiting the DRAM rowhammer bug to gain kernel privileges, Black Hat, vol. 15, 2015, p. 71
46. V. Van Der Veen, Y. Fratantonio, M. Lindorfer, D. Gruss, C. Maurice, G. Vigna, et al., Drammer: deterministic rowhammer attacks on mobile platforms. Presented at the Proceedings of the 2016 ACM SIGSAC Conference on Computer and Communications Security, Vienna, Austria, 2016
47. A. Tang, S. Sethumadhavan, S. Stolfo, CLKscrew: exposing the perils of security-oblivious energy management, in *Usenix 2018* (Distinguished Paper Award), 2018
48. M. Hutter, J.-M. Schmidt, The Temperature Side Channel and Heating Fault Attacks, Cham, 2014, pp. 219–235
49. C.H. Kim, J. Quisquater, Faults, injection methods, and fault attacks. IEEE Des. Test Comput. **24**, 544–545 (2007)
50. Y. Jin, D. Maliuk, Y. Makris, Hardware Trojan detection in analog/RF integrated circuits, in *Secure System Design and Trustable Computing*, ed. by C.-H. Chang, M. Potkonjak, (Springer International Publishing, Cham, 2016), pp. 241–268
51. Y. Liu, Y. Jin, A. Nosratinia, Y. Makris, Silicon demonstration of hardware Trojan design and detection in wireless cryptographic ICs. IEEE Trans. Very Large Scale Integr. Syst. **25**, 1506–1519 (2017)
52. K. Yang, M. Hicks, Q. Dong, T. Austin, D. Sylvester, Exploiting the analog properties of digital circuits for malicious hardware. Commun. ACM **60**, 83–91 (2017)
53. R. Kumar, P. Jovanovic, W. Burleson, I. Polian, Parametric trojans for fault-injection attacks on cryptographic hardware, in *2014 Workshop on Fault Diagnosis and Tolerance in Cryptography*, 2014, pp. 18–28
54. M. Ender, S. Ghandali, A. Moradi, C. Paar, The first thorough side-channel hardware trojan, in *International Conference on the Theory and Application of Cryptology and Information Security*, 2017, pp. 755–780
55. X.T. Ng, Z. Naj, S. Bhasin, D.B. Roy, J.-L. Danger, S. Guilley, Integrated sensor: a backdoor for hardware Trojan insertions? in *2015 Euromicro Conference on Digital System Design*, 2015, pp. 415–422
56. Y. Shiyanovskii, F. Wolff, A. Rajendran, C. Papachristou, D. Weyer, W. Clay, Process reliability based trojans through NBTI and HCI effects, in *2010 NASA/ESA Conference on Adaptive Hardware and Systems*, 2010, pp. 215–222
57. S. Bhasin, J.-L. Danger, S. Guilley, X.T. Ngo, L. Sauvage, Hardware Trojan horses in cryptographic IP cores, in *2013 Workshop on Fault Diagnosis and Tolerance in Cryptography*, 2013, pp. 15–29
58. M. Yoshimura, A. Ogita, T. Hosokawa, A smart Trojan circuit and smart attack method in AES encryption circuits, in *2013 IEEE International Symposium on Defect and Fault Tolerance in VLSI and Nanotechnology Systems (DFTS)*, 2013, pp. 278–283
59. M. Fyrbiak, S. Wallat, P. Swierczynski, M. Hoffmann, S. Hoppach, M. Wilhelm, et al., HAL—the missing piece of the puzzle for hardware reverse engineering, Trojan detection and insertion. IEEE Trans. Depend. Secure Comput. (2018)

60. L. Lin, W. Burleson, C. Paar, MOLES: malicious off-chip leakage enabled by side-channels, in *2009 IEEE/ACM International Conference on Computer-Aided Design-Digest of Technical Papers*, 2009, pp. 117–122
61. U. Guin, D. Forte, M. Tehranipoor, Design of accurate low-cost on-chip structures for protecting integrated circuits against recycling. IEEE Trans. Very Large Scale Integr. Syst. **24**, 1233–1246 (2016)
62. M.M. Alam, M. Tehranipoor, D. Forte, Recycled FPGA detection using exhaustive LUT path delay characterization and voltage scaling. IEEE Trans. Very Large Scale Integr. Syst. **27**, 2897–2910 (2019)
63. A. Stern, U. Botero, F. Rahman, D. Forte, M. Tehranipoor, EMFORCED: EM-based fingerprinting framework for remarked and cloned counterfeit IC detection using machine learning classification. IEEE Trans. Very Large Scale Integr. Syst. **28**, 363–375 (2020)
64. S. Shinde, S. Jothibasu, M.T. Ghasr, R. Zoughi, Wideband microwave reflectometry for rapid detection of dissimilar and aged ICs. IEEE Trans. Instrum. Meas. **66**, 2156–2165 (2017)
65. Y. Zheng, S. Yang, S. Bhunia, SeMIA: self-similarity-based IC integrity analysis. IEEE Trans. Comput. Aided Des. Integr. Circuits Syst. **35**, 37–48 (2016)
66. X. Wang, Y. Han, M. Tehranipoor, System-level counterfeit detection using on-chip ring oscillator array. IEEE Trans. Very Large Scale Integr. Syst. **27**, 2884–2896 (2019)
67. Z. Guo, X. Xu, M.T. Rahman, M.M. Tehranipoor, D. Forte, SCARe: an SRAM-based countermeasure against IC recycling. IEEE Trans. Very Large Scale Integr. Syst. **26**, 744–755 (2018)
68. M. Tehranipoor, U. Guin, D. Forte, Chip ID, in *Counterfeit Integrated Circuits: Detection and Avoidance*, (Springer International Publishing, Cham, 2015), pp. 243–263
69. J. Mchale, DNA marking for counterfeit parts: problem solver or money pit? *Military Embedded Systems*, 2016. Available: http://mil-embedded.com/articles/dna-problem-solver-money-pit/
70. L. Daihyun, J.W. Lee, B. Gassend, G.E. Suh, M. Van Dijk, S. Devadas, Extracting secret keys from integrated circuits. IEEE Trans. Very Large Scale Integr. Syst. **13**, 1200–1205 (2005)
71. B. Halak, Reliability challenges of silicon-based physically unclonable functions, in *Physically Unclonable Functions: From Basic Design Principles to Advanced Hardware Security Applications*, (Springer International Publishing, Cham, 2018), pp. 53–71
72. U. Rührmair, J. Sölter, F. Sehnke, X. Xu, A. Mahmoud, V. Stoyanova, et al., PUF modeling attacks on simulated and silicon data. IEEE Trans. Inf. Foren. Security **8**, 1876–1891 (2013)
73. Y. Lao, B. Yuan, C.H. Kim, K.K. Parhi, Reliable PUF-based local authentication with self-correction. IEEE Trans. Comput. Aided Des. Integr. Circuits Syst. **36**, 201–213 (2017)
74. A.E. Caldwell, C. Hyun-Jin, A.B. Kahng, S. Mantik, M. Potkonjak, Q. Gang, et al., Effective iterative techniques for fingerprinting design IP. IEEE Trans. Comput. Aided Des. Integr. Circuits Syst. **23**, 208–215 (2004)
75. C. Dunbar, Q. Gang, A practical circuit fingerprinting method utilizing observability don't care conditions, in *2015 52nd ACM/EDAC/IEEE Design Automation Conference (DAC)*, 2015, pp. 1–6
76. C. Dunbar, G. Qu, Satisfiability don't care condition based circuit fingerprinting techniques, in *The 20th Asia and South Pacific Design Automation Conference*, 2015, pp. 815–820
77. Y.M. Alkabani, F. Koushanfar, Active hardware metering for intellectual property protection and security. Presented at the Proceedings of 16th USENIX Security Symposium on USENIX Security Symposium, Boston, MA, 2007
78. A. Cui, Y. Yang, G. Qu, H. Li, A secure and low-overhead active IC metering scheme, in *2019 IEEE 37th VLSI Test Symposium (VTS)*, 2019, pp. 1–6
79. F. Koushanfar, Provably secure active IC metering techniques for piracy avoidance and digital rights management. IEEE Trans. Inf. Foren. Security **7**, 51–63 (2012)
80. J.A. Roy, F. Koushanfar, I.L. Markov, EPIC: ending piracy of integrated circuits, in *2008 Design, Automation and Test in Europe*, 2008, pp. 1069–1074

81. A. Cui, X. Qian, G. Qu, H. Li, A new active IC metering technique based on locking scan cells, in *2017 IEEE 26th Asian Test Symposium (ATS)*, 2017, pp. 40–45
82. S.M. Plaza, I.L. Markov, Solving the third-shift problem in IC piracy with test-aware logic locking. IEEE Trans. Comput. Aided Des. Integr. Circuits Syst. **34**, 961–971 (2015)
83. J.D. Rolt, G.D. Natale, M. Flottes, B. Rouzeyre, New security threats against chips containing scan chain structures, in *2011 IEEE International Symposium on Hardware-Oriented Security and Trust*, 2011, pp. 110–110
84. A. Basak, S. Bhunia, P-Val: antifuse-based package-level defense against counterfeit ICs. IEEE Trans. Comput. Aided Des. Integr. Circuits Syst. **35**, 1067–1078 (2016)
85. L. Aniello, B. Halak, P. Chai, R. Dhall, M. Mihalea, A. Wilczynski, Towards a Supply Chain Management System for Counterfeit Mitigation Using Blockchain and PUF, arXiv:1908.09585, 2019
86. X. Xu, F. Rahman, B. Shakya, A. Vassilev, D. Forte, M. Tehranipoor, Electronics supply chain integrity enabled by blockchain. ACM Trans. Design Autom. Electr. Syst. **24**, 31:1–31:25 (2019)
87. Y. Obeng, C. Nolan, D. Brown, Hardware security through chain assurance, in *2016 Design, Automation & Test in Europe Conference & Exhibition (DATE)*, 2016, pp. 1535–1537
88. C.E. Shearon, A practical way to limit counterfeits, in *2019 Pan Pacific Microelectronics Symposium (Pan Pacific)*, 2019, pp. 1–7
89. T. i. c. T. program, Intelligence Advanced Research Projects Activity, 2011
90. Y. Wang, P. Chen, J. Hu, J.J.V. Rajendran, The cat and mouse in split manufacturing, in *2016 53nd ACM/EDAC/IEEE Design Automation Conference (DAC)*, 2016, pp. 1–6
91. Y. Wang, P. Chen, J. Hu, G. Li, J. Rajendran, The cat and mouse in split manufacturing. IEEE Trans. Very Large Scale Integr. Syst. **26**, 805–817 (2018)
92. J. Rajendran, O. Sinanoglu, R. Karri, Is split manufacturing secure? in *2013 Design, Automation & Test in Europe Conference & Exhibition (DATE)*, 2013, pp. 1259–1264
93. M. Jagasivamani, P. Gadfort, M. Sika, M. Bajura, M. Fritze, Split-fabrication obfuscation: Metrics and techniques, in *2014 IEEE International Symposium on Hardware-Oriented Security and Trust (HOST)*, 2014, pp. 7–12
94. A. Sengupta, S. Patnaik, J. Knechtel, M. Ashraf, S. Garg, O. Sinanoglu, Rethinking split manufacturing: an information-theoretic approach with secure layout techniques, in *2017 IEEE/ACM International Conference on Computer-Aided Design (ICCAD)*, 2017, pp. 329–326
95. R.S. Chakraborty, S. Bhunia, Security against hardware Trojan through a novel application of design obfuscation, in *2009 IEEE/ACM International Conference on Computer-Aided Design—Digest of Technical Papers*, 2009, pp. 113–116
96. J. Dofe, Q. Yu, Novel dynamic state-deflection method for gate-level design obfuscation. IEEE Trans. Comput. Aided Des. Integr. Circuits Syst. **37**, 273–285 (2018)
97. A.R. Desai, M.S. Hsiao, C. Wang, L. Nazhandali, S. Hall, Interlocking obfuscation for anti-tamper hardware. Presented at the Proceedings of the 8th Annual Cyber Security and Information Intelligence Research Workshop, Oak Ridge, Tennessee, USA, 2013
98. M. Fyrbiak, S. Wallat, J. Déchelotte, N. Albartus, S. Böcker, R. Tessier, et al., On the difficulty of FSM-based hardware obfuscation. IACR Trans. Cryptogr. Hardware Embed. Syst. **2018**, 293–330 (2018)
99. A. Baumgarten, A. Tyagi, J. Zambreno, Preventing IC piracy using reconfigurable logic barriers. IEEE Des. Test Comput. **27**, 66–75 (2010)
100. R.S. Chakraborty, S. Bhunia, Security through obscurity: an approach for protecting register transfer level hardware IP, in *2009 IEEE International Workshop on Hardware-Oriented Security and Trust*, 2009, pp. 96–99
101. J. Rajendran, H. Zhang, C. Zhang, G.S. Rose, Y. Pino, O. Sinanoglu, et al., Fault analysis-based logic encryption. IEEE Trans. Comput. **64**, 410–424 (2015)
102. P. Subramanyan, S. Ray, S. Malik, Evaluating the security of logic encryption algorithms, in *2015 IEEE International Symposium on Hardware Oriented Security and Trust (HOST)*, 2015, pp. 137–143

103. M. Yasin, B. Mazumdar, J.J.V. Rajendran, O. Sinanoglu, SARLock: SAT attack resistant logic locking, in *2016 IEEE International Symposium on Hardware Oriented Security and Trust (HOST)*, 2016, pp. 236–241
104. Y. Xie, A. Srivastava, Anti-SAT: mitigating SAT attack on logic locking. IEEE Trans. Comput. Aided Des. Integr. Circuits Syst. **38**, 199–207 (2019)
105. L. Li, A. Orailoglu, Piercing logic locking keys through redundancy identification, in *2019 Design, Automation & Test in Europe Conference & Exhibition (DATE)*, 2019, pp. 540–545
106. R. Torrance, D. James, The state-of-the-art in semiconductor reverse engineering. Presented at the Proceedings of the 48th Design Automation Conference, San Diego, CA, 2011
107. L.W. Chow, J.P. Baukus, B.J. Wang, R.P. Cocchi, Camouflaging a standard cell based integrated circuit, Google Patents, 2012
108. R.P. Cocchi, J.P. Baukus, B.J. Wang, L.W. Chow, P. Ouyang, Building block for a secure CMOS logic cell library, Google Patents, 2012
109. M. El Massad, S. Garg, M.V. Tripunitara, Integrated circuit (IC) decamouflaging: reverse engineering camouflaged ICs within minutes, in *NDSS*, 2015, pp. 1–14
110. C. Yu, X. Zhang, D. Liu, M. Ciesielski, D. Holcomb, Incremental SAT-based reverse engineering of camouflaged logic circuits. IEEE Trans. Comput. Aided Des. Integr. Circuits Syst. **36**, 1647–1659 (2017)
111. M. Yasin, B. Mazumdar, O. Sinanoglu, J. Rajendran, Security analysis of anti-sat, in *2017 22nd Asia and South Pacific Design Automation Conference (ASP-DAC)*, 2017, pp. 342–347
112. M. Li, K. Shamsi, T. Meade, Z. Zhao, B. Yu, Y. Jin, et al., Provably secure camouflaging strategy for IC protection. IEEE Trans. Comput. Aided Des. Integr. Circuits Syst. **38**, 1399–1412 (2019)
113. A.T. Abdel-Hamid, S. Tahar, A. El Mostapha, IP watermarking techniques: survey and comparison, in *Proceedings of the 3rd IEEE International Workshop on System-on-Chip for Real-Time Applications, 2003*, 2003, pp. 60–65
114. Newcastle University, Cryptographic processing and processors, U.K. Patent Appl. No. 0719455.8, 4 Oct 2007
115. B. Halak, J. Murphy, A. Yakovlev, Power balanced circuits for leakage-power-attacks resilient design, in *Science and Information Conference (SAI), 2015*, 2015, pp. 1178–1183
116. D. Bellizia, G. Scotti, A. Trifiletti, TEL logic style as a countermeasure against side-channel attacks: secure cells library in 65nm CMOS and experimental results. IEEE Trans. Circuits Syst. I Regular Papers **65**, 3874–3884 (2018)
117. J. Shen, L. Geng, F. Zhang, Dynamic current mode logic based flip-flop design for robust and low-power security integrated circuits. Electron. Lett. **53**, 1236–1238 (2017)
118. Y. Bi, K. Shamsi, J. Yuan, Y. Jin, M. Niemier, X.S. Hu, Tunnel FET current mode logic for DPA-resilient circuit designs. IEEE Trans. Emerg. Top. Comput. **5**, 340–352 (2017)
119. M. Avital, H. Dagan, O. Keren, A. Fish, Randomized multitopology logic against differential power analysis. IEEE Trans. Very Large Scale Integr. Syst. **23**, 702–711 (2015)
120. K. Chong, A. Shreedhar, N.K.Z. Lwin, N.A. Kyaw, W. Ho, C. Wang, et al., Side-channel-attack resistant dual-rail asynchronous-logic AES accelerator based on standard library cells, in *2019 Asian Hardware Oriented Security and Trust Symposium (AsianHOST)*, 2019, pp. 1–7
121. W. Yu, S. Köse, Exploiting voltage regulators to enhance various power attack countermeasures. IEEE Trans. Emerg. Top. Comput. **6**, 244–257 (2018)
122. M. Kar, A. Singh, S.K. Mathew, A. Rajan, V. De, S. Mukhopadhyay, Reducing power side-channel information leakage of AES engines using fully integrated inductive voltage regulator. IEEE J. Solid State Circuits **53**, 2399–2414 (2018)
123. A. Gornik, A. Moradi, J. Oehm, C. Paar, A hardware-based countermeasure to reduce side-channel leakage: design, implementation, and evaluation. IEEE Trans. Comput. Aided Des. Integr. Circuits Syst. **34**, 1308–1319 (2015)
124. D. Das, S. Maity, S.B. Nasir, S. Ghosh, A. Raychowdhury, S. Sen, ASNI: attenuated signature noise injection for low-overhead power side-channel attack immunity. IEEE Trans. Circuits Syst. I Regular Papers **65**, 3300–3311 (2018)

125. M.L. Akkar, Power analysis, what is now possible, in *ASIACRYPT*, 2000
126. S. Tillich, C. Herbst, S. Mangard, Protecting AES software implementations on 32-bit processors against power analysis, in *Applied Cryptography and Network Security*, (Heidelberg, Berlin, 2007), pp. 141–157
127. C. Clavier, J.S. Coron, N. Dabbous, Differential power analysis in the presence of hardware countermeasures, in *Proceedings of the 2nd International Workshop on Cryptographic Hardware and Embedded Systems*, LNCS, vol. 1965, 2000, pp. 252–263
128. S. Tillich, J. Großschädl, Power analysis resistant AES implementation with instruction set extensions, in *Cryptographic Hardware and Embedded Systems—CHES 2007*, vol. 2007, (Heidelberg, Berlin), pp. 303–319
129. A.G. Bayrak, F. Regazzoni, D. Novo, P. Brisk, F. Standaert, P. Ienne, Automatic application of power analysis countermeasures. IEEE Trans. Comput. **64**, 329–341 (2015)
130. J. Blömer, J. Guajardo, V. Krummel, Provably secure masking of AES, in *Selected Areas in Cryptography*, (Heidelberg, Berlin, 2005), pp. 69–83
131. M. Medwed, F.-X. Standaert, J. Großschädl, F. Regazzoni, Fresh re-keying: security against side-channel and fault attacks for low-cost devices, in *Progress in Cryptology—AFRICACRYPT 2010*, (Heidelberg, Berlin, 2010), pp. 279–296
132. H. Lee, C.S. Juvekar, J. Kwong, A.P. Chandrakasan, A nonvolatile flip-flop-enabled cryptographic wireless authentication tag with per-query key update and power-glitch attack countermeasures. IEEE J. Solid State Circuits **52**, 272–283 (2017)
133. O. Goldreich, R. Ostrovsky, Software protection and simulation on oblivious RAMs. J. ACM (JACM) **43**, 431–473 (1996)
134. E. Brickell, G. Graunke, M. Neve, J.-P. Seifert, Software mitigations to hedge AES against cache-based software side channel vulnerabilities, *IACR Cryptology ePrint Archive*, vol. 2006, 2006
135. M. Godfrey, M. Zulkernine, A server-side solution to cache-based side-channel attacks in the cloud, in *2013 IEEE Sixth International Conference on Cloud Computing*, 2013, pp. 163–170
136. V.K. Base, Security considerations and disallowing inter-virtual machine transparent page sharing, *VMware Knowledge Base*, vol. 2080735, 2014
137. C. Percival, Cache missing for fun and profit, in *BSDCan*, 2005
138. M.M. Godfrey, M. Zulkernine, Preventing cache-based side-channel attacks in a cloud environment. IEEE Trans. Cloud Comput. **2**, 395–408 (2014)
139. Y. Mathieu, Cache-timing attacks still threaten IoT devices, in *Codes, Cryptology and Information Security: Third International Conference, C2SI 2019*, Rabat, Morocco, April 22–24, 2019, Proceedings-In Honor of Said El Hajji, 2019, p. 13
140. O. Acıiçmez, W. Schindler, Ç.K. Koç, Cache based remote timing attack on the AES, in *Cryptographers' Track at the RSA Conference*, 2007, pp. 271–286
141. Intel, Intel Analysis of Speculative Execution Side Channels, 2018
142. R.B. Ke Sun, K. Hu, A new memory type against speculative side channel attacks, Intel—STrategic Offensive Research & Mitigations (STORM), 2019
143. J. Fustos, F. Farshchi, H. Yun, SpectreGuard: an efficient data-centric defense mechanism against spectre attacks, in *2019 56th ACM/IEEE Design Automation Conference (DAC)*, 2019, pp. 1–6
144. K. Barber, A. Bacha, L. Zhou, Y. Zhang, R. Teodorescu, Isolating speculative data to prevent transient execution attacks. IEEE Comput. Archit. Lett. **18**, 178–181 (2019)
145. P. Li, L. Zhao, R. Hou, L. Zhang, D. Meng, Conditional speculation: an effective approach to safeguard out-of-order execution against spectre attacks, in *2019 IEEE International Symposium on High Performance Computer Architecture (HPCA)*, 2019, pp. 264–276
146. V. Kiriansky, I. Lebedev, S. Amarasinghe, S. Devadas, J. Emer, DAWG: a defense against cache timing attacks in speculative execution processors, in *2018 51st Annual IEEE/ACM International Symposium on Microarchitecture (MICRO)*, 2018, pp. 974–987
147. C. Li, J. Gaudiot, Online detection of spectre attacks using microarchitectural traces from performance counters, in *2018 30th International Symposium on Computer Architecture and High Performance Computing (SBAC-PAD)*, 2018, pp. 25–28

148. G.T. Becker, On the pitfalls of using arbiter-PUFs as building blocks. IEEE Trans. Comput. Aided Des. Integr. Circuits Syst. **34**, 1295–1307 (2015)
149. B. Gassend, D. Clarke, M. Van Dijk, S. Devadas, Controlled physical random functions, in *18th Annual Computer Security Applications Conference, 2002. Proceedings*, 2002, pp. 149–160
150. M.S. Mispan, B. Halak, Z. Chen, M. Zwolinski, TCO-PUF: a subthreshold physical unclonable function, in *2015 11th Conference on PhD Research in Microelectronics and Electronics (PRIME)*, 2015, pp. 105–108
151. Q. Ma, C. Gu, N. Hanley, C. Wang, W. Liu, M.O. Neill, A machine learning attack resistant multi-PUF design on FPGA, in *2018 23rd Asia and South Pacific Design Automation Conference (ASP-DAC)*, 2018, pp. 97–104
152. H. Su, M. Zwolinski, B. Halak, A machine learning attacks resistant two stage physical unclonable functions design, in *2018 IEEE 3rd International Verification and Security Workshop (IVSW)*, 2018, pp. 52–55
153. D.P.S.P.H. Nguyen, C. Jin, K. Mahmood, U. Rührmair, M. van Dijk, The Interpose PUF: Secure PUF Design Against State-of-the-Art Machine Learning Attacks, 2018
154. L. MacKinnon, L. Bacon, D. Gan, G. Loukas, D. Chadwick, D. Frangiskatos, Chapter 20—Cyber security countermeasures to combat cyber terrorism, in *Strategic Intelligence Management*, ed. by B. Akhgar, S. Yates, (Butterworth-Heinemann, Oxford, 2013), pp. 234–257
155. M. Mantere, I. Uusitalo, M. Sailio, S. Noponen, Challenges of machine learning based monitoring for industrial control system networks, in *2012 26th International Conference on Advanced Information Networking and Applications Workshops*, 2012, pp. 968–972
156. R. Cockell, B. Halak, On the design and analysis of a biometric authentication system using keystroke dynamics. Cryptography **4**, 12 (2020)
157. L. Cojocar, K. Razavi, C. Giuffrida, H. Bos, Exploiting correcting codes: on the effectiveness of ecc memory against rowhammer attacks, in *2019 IEEE Symposium on Security and Privacy (SP)*, 2019, pp. 55–71
158. A. Barenghi, L. Breveglieri, I. Koren, G. Pelosi, F. Regazzoni, Countermeasures against fault attacks on software implemented AES: effectiveness and cost, in *Proceedings of the 5th Workshop on Embedded Systems Security*, 2010, pp. 1–10
159. K. Xiao, D. Forte, M. Tehranipoor, A novel built-in self-authentication technique to prevent inserting hardware trojans. IEEE Trans. Comput. Aided Des. Integr. Circuits Syst. **33**, 1778–1791 (2014)
160. S.C. Konigsmark, D. Chen, M.D. Wong, Information dispersion for trojan defense through high-level synthesis, in *Proceedings of the 53rd Annual Design Automation Conference*, 2016, pp. 1–6
161. X.T. Ngo, S. Guilley, S. Bhasin, J.-L. Danger, Z. Najm, Encoding the state of integrated circuits: a proactive and reactive protection against hardware trojans horses, in *Proceedings of the 9th Workshop on Embedded Systems Security*, 2014, pp. 1–10
162. X.T. Ngo, S. Bhasin, J.-L. Danger, S. Guilley, Z. Najm, Linear complementary dual code improvement to strengthen encoded circuit against hardware Trojan horses, in *2015 IEEE International Symposium on Hardware Oriented Security and Trust (HOST)*, 2015, pp. 82–87
163. M. Hicks, M. Finnicum, S.T. King, M.M. Martin, J.M. Smith, Overcoming an untrusted computing base: Detecting and removing malicious hardware automatically, in *2010 IEEE Symposium on Security and Privacy*, 2010, pp. 159–172
164. A. Waksman, M. Suozzo, S. Sethumadhavan, FANCI: identification of stealthy malicious logic using boolean functional analysis, in *Proceedings of the 2013 ACM SIGSAC Conference on Computer & Communications Security*, 2013, pp. 697–708
165. J. Zhang, F. Yuan, L. Wei, Y. Liu, Q. Xu, VeriTrust: verification for hardware trust. IEEE Trans. Comput. Aided Des. Integr. Circuits Syst. **34**, 1148 1161 (2015)
166. M. Rathmair, F. Schupfer, C. Krieg, Applied formal methods for hardware Trojan detection, in *2014 IEEE International Symposium on Circuits and Systems (ISCAS)*, 2014, pp. 169–172

167. J. Rajendran, A.M. Dhandayuthapany, V. Vedula, R. Karri, Formal security verification of third party intellectual property cores for information leakage, in *2016 29th International Conference on VLSI Design and 2016 15th International Conference on Embedded Systems (VLSID)*, 2016, pp. 547–552
168. B. Zhou, R. Adato, M. Zangeneh, T. Yang, A. Uyar, B. Goldberg, et al., Detecting hardware trojans using backside optical imaging of embedded watermarks, in *2015 52nd ACM/EDAC/IEEE Design Automation Conference (DAC)*, 2015, pp. 1–6
169. R.S. Chakraborty, F. Wolff, S. Paul, C. Papachristou, S. Bhunia, MERO: a statistical approach for hardware Trojan detection, in *International Workshop on Cryptographic Hardware and Embedded Systems*, 2009, pp. 396–410
170. M. Banga, M.S. Hsiao, A novel sustained vector technique for the detection of hardware Trojans, in *2009 22nd International Conference on VLSI Design*, 2009, pp. 327–332
171. A. Amelian, S.E. Borujeni, A side-channel analysis for hardware Trojan detection based on path delay measurement. J. Circuits Syst. Comput. **27**, 1850138 (2018)
172. X. Wang, H. Salmani, M. Tehranipoor, J. Plusquellic, Hardware Trojan detection and isolation using current integration and localized current analysis, in *2008 IEEE International Symposium on Defect and Fault Tolerance of VLSI Systems*, 2008, pp. 87–95
173. B. Hou, C. He, L. Wang, Y. En, S. Xie, Hardware Trojan detection via current measurement: a method immune to process variation effects, in *2014 10th International Conference on Reliability, Maintainability and Safety (ICRMS)*, 2014, pp. 1039–1042
174. L. Wang, H. Xie, H. Luo, Malicious circuitry detection using transient power analysis for IC security, in *2013 International Conference on Quality, Reliability, Risk, Maintenance, and Safety Engineering (QR2MSE)*, 2013, pp. 1164–1167
175. Z. Chen, S. Guo, J. Wang, Y. Li, Z. Lu, Toward FPGA security in IoT: a new detection technique for hardware Trojans. IEEE Internet Things J. **6**, 7061–7068 (2019)
176. G. Shen, Y. Tang, S. Li, J. Chen, B. Yang, A general framework of hardware Trojan detection: two-level temperature difference based thermal map analysis, in *2017 11th IEEE International Conference on Anti-counterfeiting, Security, and Identification (ASID)*, 2017, pp. 172–178
177. M. Cozzi, J.-M. Galliere, P. Maurine, Exploiting phase information in thermal scans for stealthy Trojan detection, in *2018 21st Euromicro Conference on Digital System Design (DSD)*, 2018, pp. 573–576
178. X.T. Ngo, J.-L. Danger, S. Guilley, Z. Najm, O. Emery, Hardware property checker for run-time hardware trojan detection, in *2015 European Conference on Circuit Theory and Design (ECCTD)*, 2015, pp. 1–4
179. A. Basak, S. Bhunia, T. Tkacik, S. Ray, Security assurance for system-on-chip designs with untrusted IPs. IEEE Trans. Inf. Foren. Security **12**, 1515–1528 (2017)
180. J. Portillo, E. John, Using static hardware wrappers to thwart hardware Trojans and code bugs at runtime, in *2018 IEEE 61st International Midwest Symposium on Circuits and Systems (MWSCAS)*, 2018, pp. 1034–1037
181. L. Changlong, Z. Yiqiang, S. Yafeng, G. Xingbo, A system-on-chip bus architecture for hardware Trojan protection in security chips, in *2011 IEEE International Conference of Electron Devices and Solid-State Circuits*, 2011, pp. 1–2
182. S.C.Y. Dhanekula, Tamper detection countermeasures to deter physical attack on a security asic, United States Patent, 2011
183. J.W.T.W. Lamfalusi, Tamper protection mesh in an electronic device, United States Patent, 2013
184. J. Obermaier, V. Immler, The past, present, and future of physical security enclosures: from battery-backed monitoring to PUF-based inherent security and beyond. J. Hardware Syst. Security **2**, 289–296 (2018)
185. A. Alattar, T. Kalker, Self-destructive and dead-on-demand devices for data protection [in the spotlight]. IEEE Signal Process. Mag. **24**, 160–158 (2007)

186. Secure Devices. Available: https://it.tufts.edu/securing-devices
187. A. Barenghi, L. Breveglieri, I. Koren, D. Naccache, Fault injection attacks on cryptographic devices: theory, practice, and countermeasures. Proc. IEEE **100**, 3056–3076 (2012)
188. S. E. a. M. H. a. C. Paar, The End of Logic Locking? A Critical View on the Security of Logic Locking, 2019

# Part II
# Emerging Hardware-Based Security Attacks and Countermeasures

# Chapter 2
# A Cube Attack on a Trojan-Compromised Hardware Implementation of Ascon

**Jorge E. Duarte-Sanchez and Basel Halak**

## 2.1 Introduction

### *2.1.1 Motivation*

With the increasing complexity of modern integrated circuits, it is more and more common to use components from third parties and to outsource manufacturing processes. This can represent a security issue because untrusted third parties can have access to the design of a circuit and insert hardware Trojans to perform malicious operations, which can severely undermine the security of an electronics system. For example, the authors of [1] show how a hardware Trojan can be used to facilitate a denial of service attack. In [2], a dopant-level Trojan is used to undermine the source of randomness of an Intel's cryptographically secure RNG design.

HTs can compromise the security of encryption cores by, for example, leaking the key directly or indirectly, or by violating security assumptions that can expose other sensitive information [3–5]. A straightforward method for obtaining the key in a cryptographic core is by inserting a HT that exposes the key directly to the output of the cipher using a multiplexor. This type of HT, although effective, can be obvious and could be detected easily due to the large number of gates required for the implementation of the multiplexor. To make the HT less obvious and smaller, and therefore much harder to detect, a HT can modify other parameters of the cipher such that it is possible to obtain the key using additional processing. For example, in [6], a HT inserts faults before the MixColumn operation of the 8th or 9th round in an iterative implementation of AES to unveil the key using the Piret's algorithm for

J. E. Duarte-Sanchez · B. Halak (✉)
University of Southampton, Southampton, UK
e-mail: basel.halak@soton.ac.uk


B. Halak (ed.), *Hardware Supply Chain Security*,
https://doi.org/10.1007/978-3-030-62707-2_2

Differential Fault Attack. In another example, a HT modifies the control circuit of an iterative implementation of AES to output ciphertexts encrypted with more than 10 rounds, which allows to unveil the key by applying a post-processing on the obtained ciphertexts [7]. The security of cryptographic algorithms is evaluated by testing them against different attacks, and this process, known as cryptanalysis, allows to determine the minimum conditions of the algorithm to achieve a desired level of security [8]. In block ciphers, one important parameter is the number of rounds of substitution-permutation operations: in general, the higher number of rounds, the more secure is the algorithm [9, 10]. One example of cryptanalysis tool is the cube attack which can be used to unveil the key of stream and block ciphers [11]. Ascon is a block cipher for authenticated encryption that uses 12 rounds of permutations during its initialization phase [12]. The cube attack applied to a reduced version of Ascon with seven rounds of permutations is already unfeasible; however, the same attack on a five-round version of the algorithm can unveil the key in a reasonable amount of time [13].

The purpose of this work is to investigate the feasibility of using hardware Trojan to undermine the resilience of Ascon ciphers to the cube attack. To achieve this, we have developed two hardware Trojans, the first allows obtaining the key directly at the output of the circuit by modifying its data path and state machine. The second HT unveils the key indirectly by reducing the number of rounds of permutations of the initialization phase of the algorithm to five with a minimal modification of the circuit and then applying a cube attack.

The combined attack (HT + cube attack) was evaluated on a SoC FPGA with the Ascon cipher residing in the logic fabric and the cube attack running in the ARM processor system.

Our experiments indicate that it is possible to unveil the key in 94 s on average. The synthesis results of the designed Trojans show that the Trojan that reduces the number of rounds has lower overhead compared to the Trojan that exposes the key directly which makes its detection more difficult. Our proposed attack scenario shows how hardware Trojans can be used to facilitate the implementation of attacks on cryptographic systems that otherwise are unfeasible in normal conditions. To our best knowledge, this is the first work that uses a cube attack in combination with a hardware Trojan to perform an attack on a hardware implementation of a cryptographic algorithm. The remainder of this paper is organized as follows.

### 2.1.2 Chapter Summary

Section 2.2 reviews essential background related to this work. Section 2.3 discusses the hardware implementation of the Ascon cipher and its security vulnerabilities. Section 2.4 develops two hardware Trojan designs and their respective implementations. Section 2.5 demonstrates the feasibility of cube attack on the Trojan-compromised implementation of the Ascon cipher. Section 2.6 discusses possible techniques to further reduce the attack time and presents a number of design recommendations to mitigate against the risks of the proposed Trojans. Conclusions are drawn in Sect. 2.7.

## 2.2 Background

### *2.2.1 Hardware Trojans*

A Hardware Trojan (HT) is a modification of a circuit that performs a malicious operation to disrupt services, leak-sensitive data, or degrade the performance of a system. HTs can be inserted in the hardware description code (RTL) or netlist of an Intellectual Property module (IP) during design or system integration phase, or in the layout of an integrated circuit (IC) during the manufacturing process. HTs have typically two main components: the payload that is the circuit that performs the malicious operation, and the trigger circuit that activates the HT. Payload and trigger circuits are designed to have a minimum impact on the original circuit to make the detection of the HT more difficult. Additionally, to avoid being detected during functional test, HTs are normally triggered under conditions that rarely occur during the normal operation of the circuit but that the attacker knows so that he can perform the malicious operation. Protection against HTs can be approached using prevention and detection techniques [14]. Prevention techniques aim to make insertion of HT more difficult by obfuscating the design [15] or by filling with functional cells the empty areas of the layout of an IC which otherwise can be used to insert HTs [16, 17]. Detection techniques can be used in pre-silicon and post-silicon stages of the IC development. Pre-silicon detection techniques detect HT in the RTL or netlist of IPs by analyzing suspicious networks with low activation [18, 19] or inputs with low impact on the circuit outputs. Formal verification techniques can also detect HT by ensuring that the IP implementation represents exactly its specification [20, 21]. Post-silicon detection techniques detect HT once the IC has been manufactured, and they can be invasive or noninvasive. In invasive methods, an IC is de-packed, de-layered, and compared against a golden (reference) IC to detect changes in the layout which indicates the presence of HTs. Non-invasive techniques include functional testing and evaluation of the side-channel effects of the HT [22, 23].

### *2.2.2 Ascon Algorithm*

Ascon is an algorithm for Authenticated Encryption with Associated Data (AEAD) which was selected in 2019 in the CAESAR competition as an alternative to AES Galois/Counter Mode block ciphers (AES-GCM) for lightweight applications [24]. AEAD schemes provide confidentiality and authenticity of messages where encrypted information (payload) is used together with optional non-encrypted

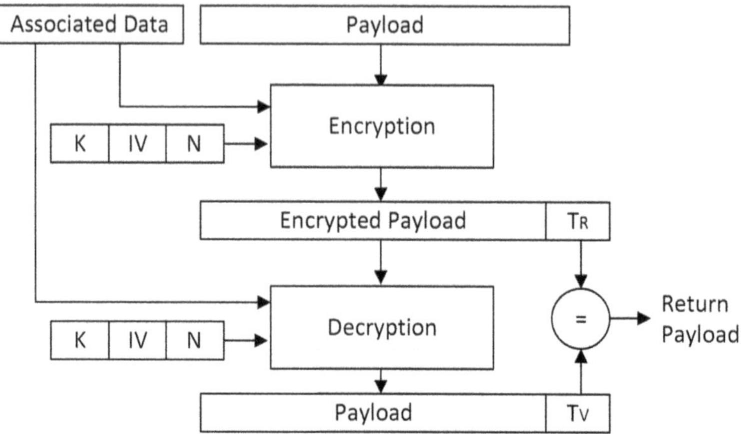


**Fig. 2.1** Authenticated encryption and decryption process with associated data using Ascon

(associated) data. For example, the associated data can correspond to a packet header which must be accessible (not encrypted) to navigate through a network, while the payload contains sensitive data that must be encrypted to keep confidentiality of the message. In an AEAD scheme, the authenticity of the associated data is achieved by embedding the data in the encryption and decryption operations. This operation is illustrated in Fig. 2.1. The payload is encrypted using the associated data and three parameters: a secret key ($K$), a constant initialization vector (IV), and a public number ($N$). $N$ works as a nonce, meaning that each encryption must use a different value of $N$. The encryption process also generates a tag reference tag ($T_{\mathrm{R}}$). To decrypt, the cipher uses the same three parameters and the associated data to recover the payload and to generate a validation tag ($T_{\mathrm{V}}$). Only if both tags are equal, the authenticity of the associated data is verified, and the cipher returns the plain payload. Ascon has two versions: Ascon-128 and Ascon-128a. In both versions, the key, nonce, and tags have 128 bits. Ascon-128 uses data blocks of 64 bits, while Ascon-128a uses blocks of 128 bits.

An Ascon encryption operation consists of four stages as illustrated in Fig. 2.2, Initialization, Associated Data, Plaintext, and Finalization [12]. The core of Ascon is a 320-bit permutation consisting on three operations: addition of constants, substitution, and linear diffusion. The Initialization and Finalization stages perform 12 rounds of permutations (operation denoted as $p^a$); The Associated data and Plaintext (payload) blocks are processed using six or eight rounds of permutations in Ascon-128 or Ascon-128a, respectively (operation denoted as $p^b$).

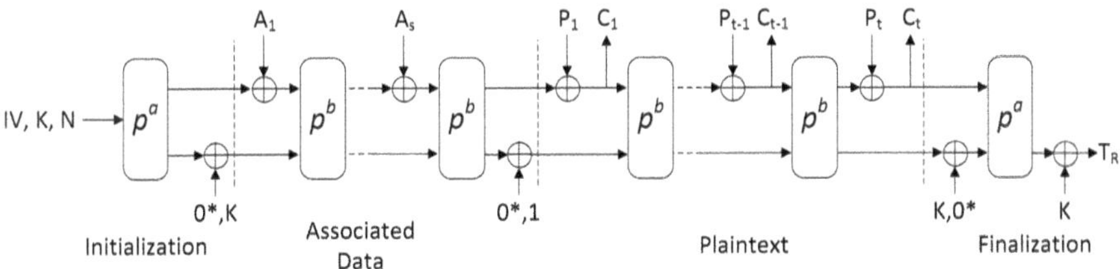


**Fig. 2.2** Mode of operation of Ascon encryption algorithm. After initialization, each block of the associated data ($A_i$) is combined with upper $r$ bits of the internal state of the cipher. Then, each block of the payload ($P_i$) is combined with upper $r$ bits of the state of the cipher to generate the blocks of encrypted payload ($C_i$). For each block of payload and associated data, the state is updated by performing six or eight rounds of permutations ($p^b$). The reference tag ($T_R$) is generated in the finalization phase combining the $k$ lower bits of the state of the system with the key ($K$)

### 2.2.3 *Cryptanalysis of Ascon*

The security of Ascon relies, among other factors, on the number of rounds of permutation operations. A security analysis of the algorithm is presented in [13] using cube-like, differential, and linear cryptanalysis. In that work, the authors describe theoretical and practical key-recovery attacks of round-reduced versions of Ascon. Furthermore, in [25] Li et al. present improvements to the cube-like cryptanalysis of Ascon to achieve a practical key-recovery attack of a reduced initialization phase of six and five rounds with time complexity of $2^{40}$ and $2^{24}$, respectively.

### 2.2.4 *Cube Attack*

The cube attack is a cryptanalysis method applicable to any stream cipher or block cipher (e.g., Ascon) [11]. The attack targets the initialization phase of the algorithm by processing ($N$, $P \oplus C$) pairs without associated data as depicted in Fig. 2.3. The basic idea of the cube attack is to recover the key one bit at the time by manipulating specific bits of $N$, called cube variables. For each key bit $k_i$, a set of cube variables is defined, and different values of $N$ are generated by setting the cube variables with all their possible binary combinations. These values of $N$ are used to encrypt an empty message ($P = 0$). The resultant encrypted messages ($C$) form a system of linear equations for $k_i$ that can be solved to obtain its value. After evaluating all the cube variables for all the key bits. There might be remaining bits of the key that cannot be obtained. Experimental results show that the number of remaining bits is always less than $2^{14}$, therefore it is feasible to find them by exhaustive search [25].

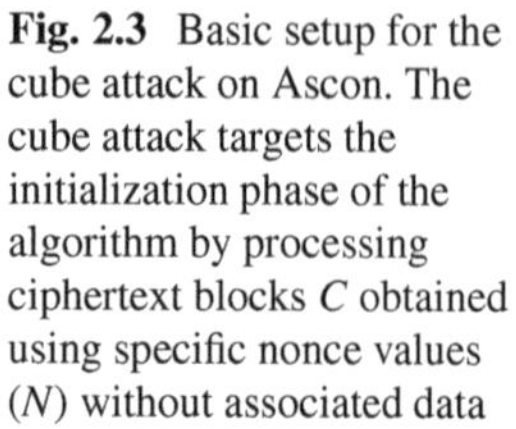

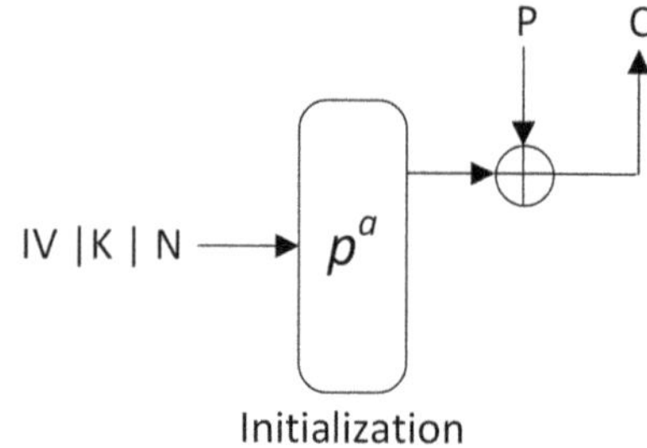


**Fig. 2.3** Basic setup for the cube attack on Ascon. The cube attack targets the initialization phase of the algorithm by processing ciphertext blocks $C$ obtained using specific nonce values ($N$) without associated data

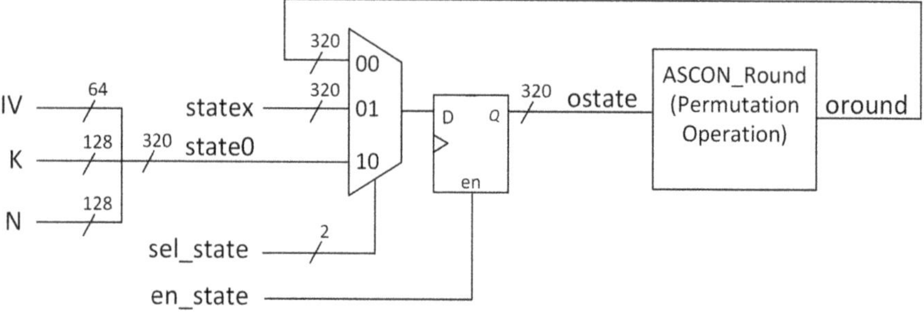


**Fig. 2.4** Schematic diagram of the processing unit of Ascon-128 CipherCore

## 2.3 Hardware Implementation of Ascon

In this work we used the CAESAR Hardware API reference implementation available in [26] to identify features of the design which allowed us to insert a hardware Trojan.

### *2.3.1 Datapath*

The main module of the implementation is the Cipher core, which is the implementation of the Ascon cipher itself. The CipherCore has an iterative implementation; this means that the same hardware required to process one round of permutation is reused as many times as the number of rounds is required for each stage of encryption or decryption. Figure 2.4 shows a schematic diagram of the Ascon CipherCore.

The algorithmic state of the system (*ostate*) is controlled using a 320-bit 3-to-1 multiplexor. In the initialization phase (*sel_state* = 10), *ostate* takes the value of *state0* which contains the initialization vector constant (IV), the key, and the nonce ($N$). Then, the block *ASCON_Round* performs one round of the permutation operation and puts the result in the signal *oround*. The multiplexor then selects *oround* as the new value of *ostate* (*sel_state* = 00) to calculate the next round

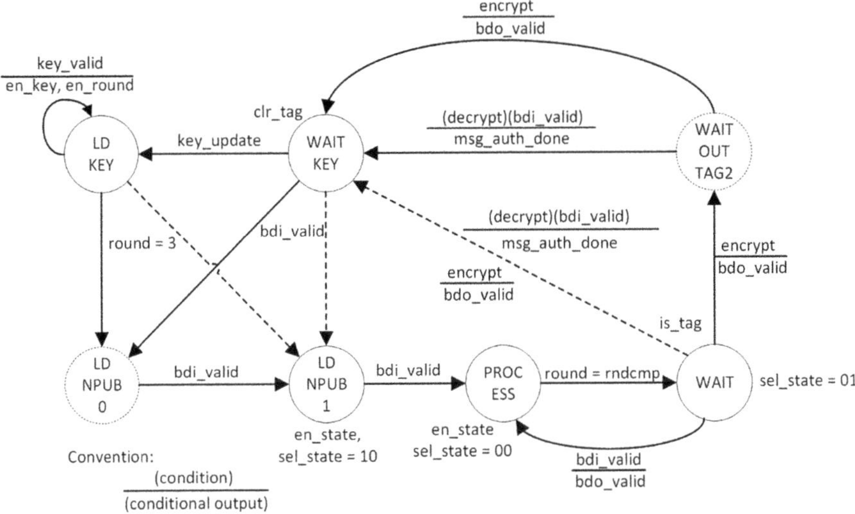


**Fig. 2.5** A simplified FSM diagram of the control unit of the CipherCore. Dashed lines represent transitions that only occur in the 128-bit version of Ascon because in this case, loading the nonce (Npub) and the reference tag takes only one clock cycle, while in the 64-bit version, it takes two

of permutation. This process is repeated until completing 12 rounds. When the CipherCore is processing ciphertext, plaintext or associated data blocks, *ostate* is updated with the value of *statex* (*sel_state* = 01) which contains the value of the current data block, and then performs six rounds of permutations.

### 2.3.2 *State Machine*

Figure 2.5 shows the FSM (Finite State Machine) diagram of the control unit of the CipherCore. The initial state is WAIT_KEY from where the key can be updated, or a new encryption or decryption operation can be started. In states LD_NPUB0 and LD_NPUB1, the nonce $N$ (Npub) is loaded (LD_NPUB0 is only used in Ascon-128 of the cipher to load the 64 most significant bits of the nonce). PROCESS state is where the permutation rounds are executed. In this state, the value of a counter (*round*) is incremented to control the number of rounds (a round is executed in one clock cycle). The condition $round = rndcmp$ indicates that the cipher has executed the necessary number of rounds for the current phase (*rndcmp* is set to 12 for the initialization and finalization phases, and to 6 for the other stages). WAIT state is used to wait for the arrival of the next block of data (associated data, plaintext, ciphertext, or $T_V$). WAIT_OUT_TAG2 is only used in Ascon-128 to load the 64 most significant bits of $T_K$. During decryption, and at the end of the finalization stage, in the state WAIT or WAIT_OUT_TAG2, $T_V$ is compared against $T_R$ to evaluate the authenticity of the associated data.

### *2.3.3 Vulnerabilities of Hardware Implementation*

1. Incomplete specification of states

   Figure 2.5 shows that the FSM of the control unit of Ascon uses a maximum of seven states. If the FSM is implemented using binary or gray encoding, it is necessary to use three flip-flops (registers) to implement it, which results in a total of eight possible states. The remaining unused states can be used to insert additional behavior to the system in the form of a hardware Trojan to perform a malicious functionality without increasing the number of registers in the circuit.
2. Unused values of primary inputs

   The unused combinations of primary inputs that do not have a specific function can be used as trigger conditions for a HT. The implementation of Ascon includes a wrapper module that implements a hardware API to evaluate all the submissions of the CAESAR competition using the same hardware interface. In the wrapper, the signal *bdi_type* identifies the type of data that is received at the input port (associated data, data, tag, nonce, etc.). However, in the CipherCore entity, *bdi_type* is only used to distinguish between associated data (*bdi_type* = 00*X*) and data (plaintext or ciphertext blocks) (*bdi_type* = 01*X*) (*X* denotes either 1 or 0). This is because other control signals determine the behavior of the cipher and the type of data that is expected in the input at each stage of operation. Therefore, since the condition *bdi_type* = 111 is never used during normal operation, it can be used as a trigger condition for a HT.
3. Iterative implementation

   In a full unrolled or pipeline implementation of Ascon, the hardware structure is fixed for a given number of permutation rounds [27]. However, in the iterative implementation of Ascon, the number of permutation rounds is controlled using a counter whose output is compared against a constant value equivalent to the required number of rounds. Therefore, by modifying only this part of the circuit, it is possible to reduce the number of rounds and void the security of the cipher.

## 2.4 Inserting Trojans in the Hardware Implementation of Ascon

In this section, we present the design of two HTs to obtain the key of the hardware implementation of Ascon. The first Trojan, called Key-Bypass Trojan, allows to obtain the key of the cipher directly by adding a malicious state in the FSM to alter the control signals such that the key can be bypassed from the key register to the output of the cipher.

The second Trojan, called Round-Reduction Trojan, reduces the number of rounds of the permutation operation during the initialization phase of encryption from 12 to 5 rounds so that it is possible to perform a cube attack with a time complexity of $2^{24}$. The description of the Trojans in the following sections refers to Ascon-128; however, the Trojans can be used also for Ascon-128a.

### 2.4.1 *Key-Bypass Trojan*

The unused states of the FSM and the incomplete inputs of the multiplexors can be used to design a Trojan that allows getting the key of the Cipher core. Figure 2.6 illustrates how to get the key of the Cipher by including a fourth input to the 3-1 multiplexor of the Cipher's data path. The additional output is activated in a malicious state inserted in the state machine as illustrated in Fig. 2.7.

1. Trojan payload

   In order to bypass the key, four signals of the FSM must be modified: *sel_state*, *en_state*, *bdo_valid*, and *en_round*. The value of these signals depends on the current state and, for some of them, on the inputs of the FSM. Figure 2.8 shows the modifications required to implement the Trojan for the signal *bdo_valid* (similar modifications are necessary for *en_round*). The modifications assume an FSM implemented with a binary state encoding as shown in Table 2.1. Without the Trojan, *bdo_valid* is activated in states *WAIT* ($S_2S_1S_0 = 101$) and *WAIT_TAG2* ($S_2S_1S_0 = 110$) when other conditions are also met. To implement the Trojan, it is necessary that *bdo_valid* is also activated in state *TROJAN*. This

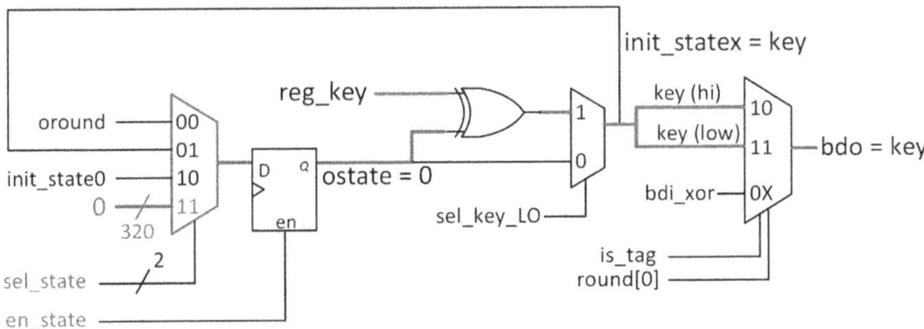


**Fig. 2.6** Trojan to bypass the key to the output of the CipherCore. A fourth input containing only 0s is added to the multiplexor that selects the value of *ostate*. This input is selected at the end of the finalization phase of encryption when the tag is being generated (*is_tag* = 1) by setting *sel_state* = 11 in a new state (TROJAN) so that the key is XORed with 0s, which effectively bypass the key to the output. If the ostate register has a clear or reset input, this trojan can also be implemented by clearing the *ostate* register when the trojan is triggered

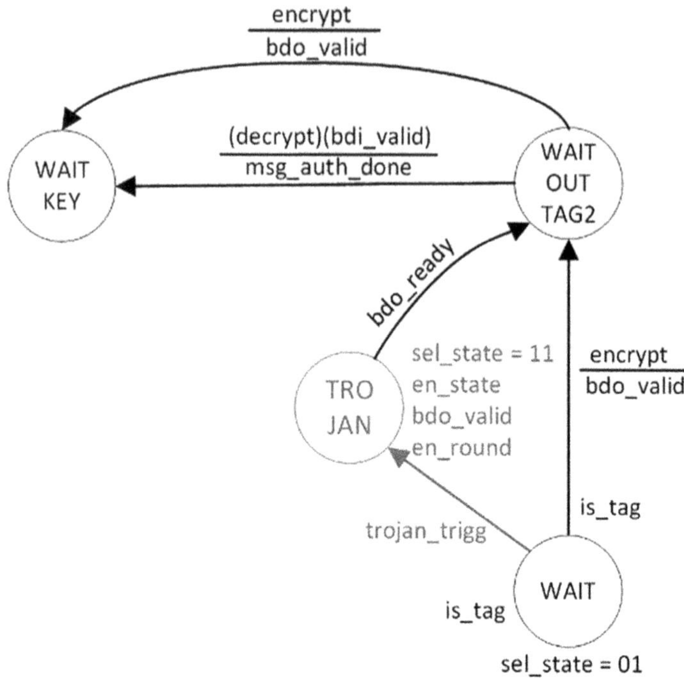


**Fig. 2.7** Insertion of the state TROJAN in the FSM to bypass the key to the output of the CipherCore

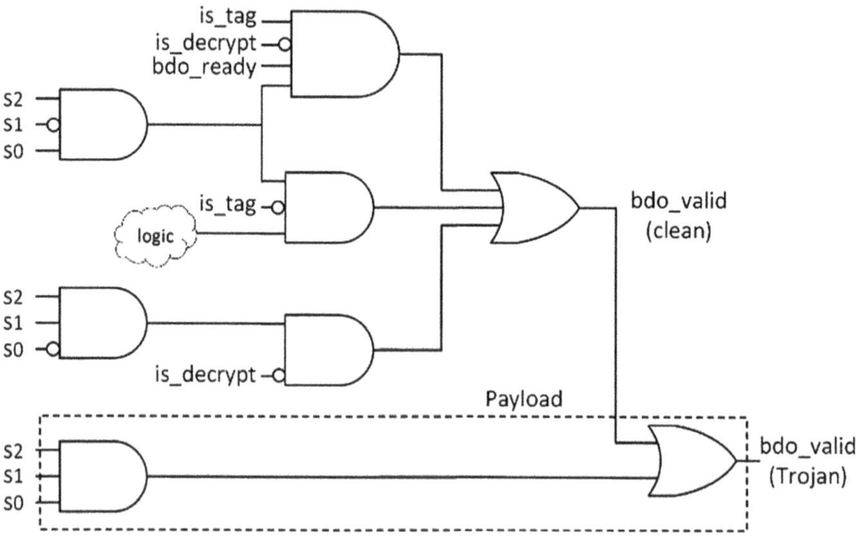


**Fig. 2.8** Modifications to signal *bdo_valid* to implement the Key-Bypass Trojan

**Table 2.1** Assumed state codifications of the FSM of the CipherCore

| State | Codification | | |
|---|---|---|---|
| | $S_2$ | $S_1$ | $S_0$ |
| WAIT_KEY | 0 | 0 | 0 |
| LD_KEY | 0 | 0 | 1 |
| LD_NPUB0 | 0 | 1 | 0 |
| LD_NPUB1 | 0 | 1 | 1 |
| PROCESS | 1 | 0 | 0 |
| WAIT | 1 | 0 | 1 |
| WAIT_TAG2 | 1 | 1 | 0 |
| TROJAN | 1 | 1 | 1 |

is achieved by introducing an AND gate and an OR gate as shown in the lower part of Fig. 2.8. Without the Trojan, *sel_state*[0] is active only in state *WAIT* ($S_2S_1S_0 = 101$), and with the Trojan, *sel_state*[0] must also be activated in state *TROJAN* ($S_2S_1S_0 = 111$). As shown in Fig. 2.9a, the minterms of the two states can be combined either to reduce the logic of *sel_state*[0] to a two-input AND gate or to connect the middle input to 0. A similar situation occurs with *sel_state*[1] as shown in Fig. 2.9b and with the signal *en_state*. In order to insert the state *TROJAN*, it is necessary to modify the logic that generates the next value for each state variable of the FSM ($S_2$, $S_1$, and $S_0$). The next value of a state variable Si is determined by the current state (current value of $S_2S_1S_0$) and by additional logic which depends on the FSM inputs (control signals). Figure 2.10 shows the specific modifications required to insert the state *TROJAN*. For the state variable $S_0$, one AND gate must be inserted to enable the transition from state *WAIT* to state *TROJAN* when the Trojan is triggered (*t_trigg* = 1). For $S_1$ and $S_2$, it is necessary to insert one AND gate that evaluates the state codification of the state *TROJAN*, and another AND gate with *bdo_ready* signal to control the transition to the state *WAIT_OUT_TAG2*. Additionally, for each state variable ($S_0$–$S_1$), it is also necessary to insert one OR gate between the existing OR gates and the input of the state registers.

2. Trojan trigger

   As explained in Sect. 2.3.3, the unused input value *bdi_type* = 111 can be used as trigger condition for the Trojan. This can be implemented with a three-input AND gate as shown in Fig. 2.10a.

## 2.4.2 *Round-Reduction Trojan*

As shown in Fig. 2.11, The number of rounds of permutation in the Cipher core is determined by two constants: TOT_RND_HI and TOT_RND_LO. TOT_RND_HI, which is equal to 12, is used to set the number of rounds in the initialization and finalization phases. TOT_RND_LO, which is equal to 6, is used to set the number

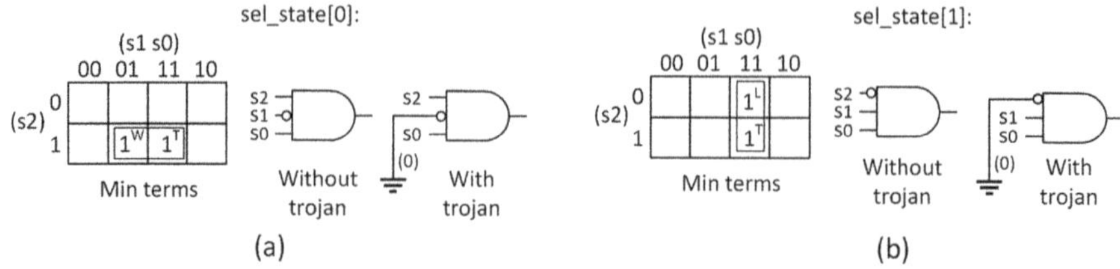


**Fig. 2.9** Modifications to signal *sel_state* to implement the Trojan to bypass the key. (**a**) Modifications to signal *sel_state*[0]. (**b**) Modifications to signal *sel_state*[1]

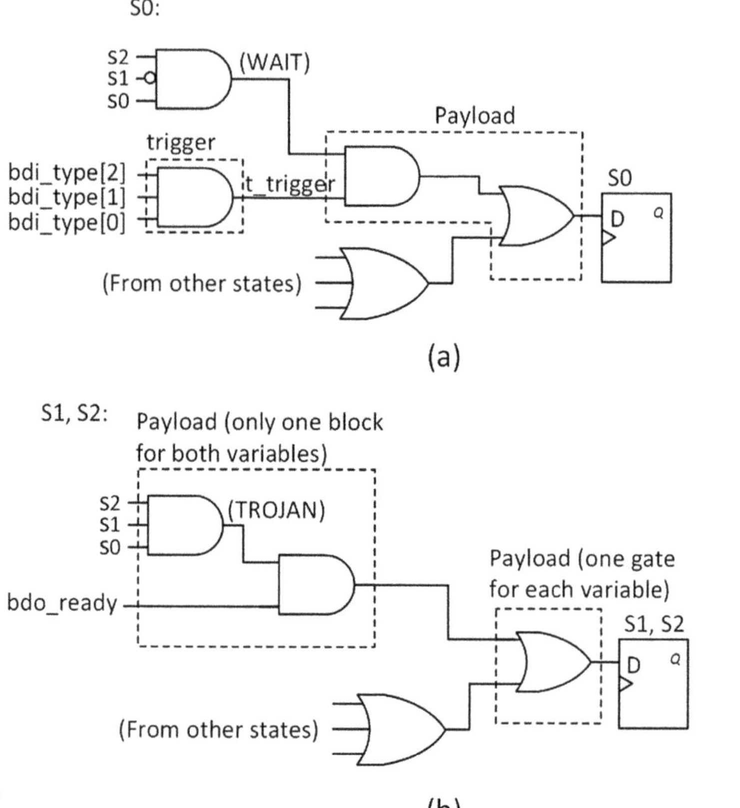


**Fig. 2.10** Required modifications to insert the state TROJAN to the FSM of the control unit of the CipherCore. (**a**) Payload and trigger logic required for $S_0$. (**b**) Payload required for $S_1$ and $S_2$

of rounds when the cipher is processing the associated data, the plaintext, and the ciphertext. *set_compute_hi* and *set_compute_lo* are control signals used to set the value of the register *rndcmp* to TOT_RND_HI or TOT_RND_LO depending on the current state of the system. When the value of the counter *round* is equal to the value

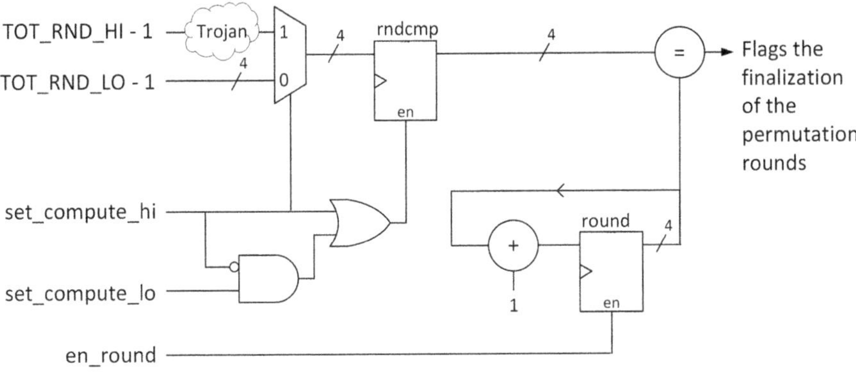


**Fig. 2.11** Trojan insertion in the control circuit to reduce the number of rounds of permutations in Ascon

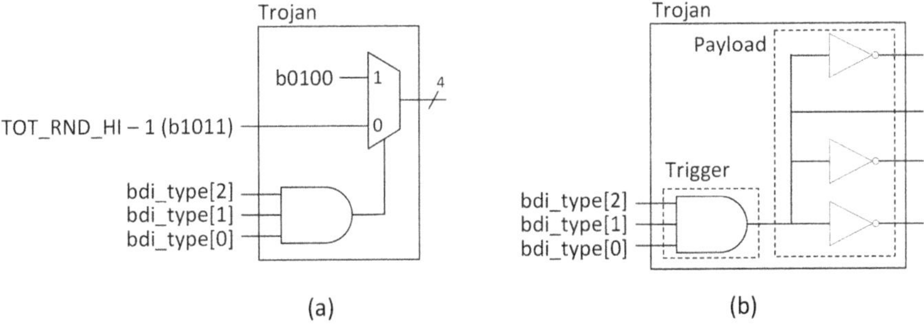


**Fig. 2.12** Trojan for round reduction. (**a**) Model of the Trojan with a multiplexor and one AND gate. (**b**) Equivalent circuit of the Trojan replacing the multiplexor with NOT gates and wires

of *rndcmp*, the cipher stops processing rounds of permutations and waits until a new data block arrives. By modifying the value of these constants, it is possible to reduce the number of rounds.

1. Trojan payload
   The number of rounds of the initialization phase can be changed by inserting a 4-bit multiplexer that selects between the normal expected value TOT_RND_HI – 1 (11), and the desired value, which in this case is 4 in order to execute five rounds of permutation as shown in Fig. 2.12a. Since the inputs of the multiplexor are constants and complementary, the same operation can be simplified and implemented using only NOT gates and wires as it is shown in Fig. 2.12b. As a result, the designed HT has a small footprint, which makes it relatively easy to insert in the layout and hard to detect.
2. Trojan trigger
   The condition *bdi_type* = 111 can also be used as a trigger condition for the Round-Reduction Trojan.

## 2.5 Cube Attack on a Trojan-Compromised Hardware Implementation of Ascon

A theoretical cube attack on Ascon exists for an implementation of the algorithm with an initialization phase consisting of maximum seven rounds of permutation [13]. This attack has a time complexity of $2^{103.9}$ which is not practical. However, if the number of rounds is further reduced to 5, for example, by inserting the Round-Reduce Trojan described previously, it is possible to perform a cube attack with a time complexity of $2^{24}$. In this section, we present a practical attack that uses the Round-Reduction Trojan and the cube attack to obtain the key of a hardware implementation of Ascon in a SoC FPGA.

### *2.5.1 Attack Assumptions*

In order to reduce the number of rounds of the initialization phase, the attacker must be able to insert the HT in either the RTL or netlist during the integration phase of the design of a SoC, or in the layout of the circuit before manufacturing.

With the infected cipher already deployed in the field, the attacker must have access to the system that uses the cipher to insert a malware (malicious software) to activate the HT and perform the cube attack. To activate the HT, the malware must have the capability of altering the function that sends the nonce ($N$) to the core so that it writes "111" to *bdi_type* when the nonce value is sent to activate the Trojan. The malware runs along with other processes in a computing system with an operating system (OS). To perform the attack, the malware must perform encryption operations while the other applications are not using the cipher. Every time that the operating system grants execution time to the malware, it activates the HT and runs a part of the attack performing as many encryptions as possible before the OS pre-empts and executes a diffident process. The malware uses the obtained cipher texts Cn for each nonce value Nn to solve the system of linear equations and find the key.

### *2.5.2 Experimental Evaluation*

1. Setup

   To determine the feasibility of the proposed attack (round-reduction HT + cube attack), the attack was evaluated in a development board DE1-SoC with a Cyclone V SoC FPGA (5CSEMA5F31C6). Figure 2.13 shows a block diagram of the experimental setup. The Ascon cipher (CipherCore) was implemented in the logic fabric of the FPGA, and the code to perform the cube attack was executed in the ARM processor embedded in the chip.

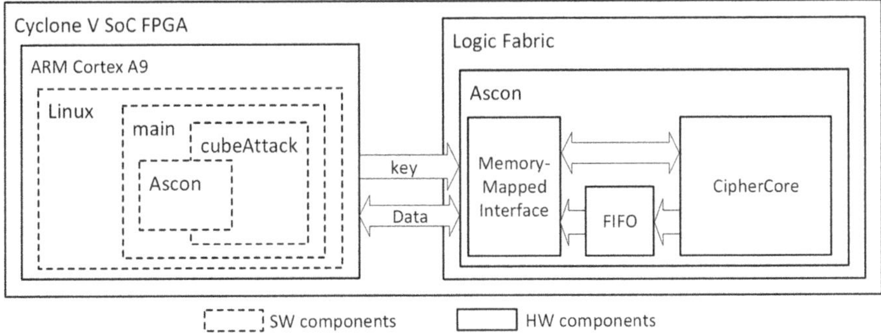


**Fig. 2.13** Setup to perform the cube attack on the Hardware implementation of Ascon

**Table 2.2** Summary of the results of 50 cube attacks

| Metric | Min. | Avg. | Max. |
|---|---|---|---|
| Cube attack encryptions | 11,927,552 | 12,702,188 | 13,762,560 |
| Remaining key bits | 0 | 3.3 | 12 |
| Exhaustive search encryptions | 0 | 16.2 | 148 |
| Total attack time (s) | 89 | 94.36 | 103 |

The code for the cube attack is based on the one used in [25] which is available in [27]. The code was modified to encrypt using the CipherCore through an AXI4 Lite memory-mapped interface, instead of using the software functions of the original code.

2. Results

Table 2.2 summarizes the results of 50 cube attacks using random keys after activating the designed Trojan in the CipherCore. The table shows the minimum, average, and maximum of the number of the encryption operations required to perform the cube attack, the number of remaining key bits that were not guessed after performing the cube attack, the number of encryptions required to find the remaining bits by exhaustive search, and the total attack time. In order to evaluate the hardware overhead and the performance impact of the designed Trojan, the CipherCore was synthesized for FPGA and ASIC targets with and without the Trojan. A summary of the synthesis results is shown in Table 2.3.

### 2.5.3 *Trojans Overhead*

The results shown in Table 2.3 confirm that the designed trojans have a low overhead; however, these results do not represent precisely the effect of the Trojans. For example, the results show that in FPGA the Round-Reduction Trojan decreases

**Table 2.3** Synthesis results of the CipherCore without and with the designed Trojans

| Target | Parameter | Clean | Round-reduction | Key-bypass |
|---|---|---|---|---|
| ASIC | Combinational cells | 3983 | 3971 | 4024 |
| | Registers | 535 | 536 | 535 |
| | Buf/Inv | 885 | 869 | 732 |
| | Total dynamic power (mW) | 22.5 | 22.3 | 22.4 |
| | Maximum path delay (ns) | 11.5 | 11.5 | 11.5 |
| FPGA | ALMs | 771 | 771 | 827 |
| | Registers | 534 | 535 | 534 |
| | Maximum path delay (ns) | 4.8 | 4.75 | 4.9 |

the maximum path delay of the circuit, but since this trojan is not located in the critical path of the original or the modified circuit, this parameter should not change. This can be explained considering that small modifications in the design can result in different logic optimizations, resource utilization, routing, and critical paths because the initial conditions (seed values) of the synthesis algorithm change.

Nevertheless, the results show that, in both FPGA and ASIC, the circuit with Round-Reduction Trojan uses one more register than the original circuit. This is because without the trojan, the register rndcmp[0] is optimized away since the two values that it can take, TOT_RND_HI[0] or TOT_RND_LO[0], are equal to 1. On the other hand, with the trojan, the value of TOT_RND_HI can be even (5-1) or odd (12-1), and therefore, TOT_RND_HI[0] can be 0 or 1, which prevents rndcmp[0] to be optimized away. This can be avoided if the Trojan reduces the number of rounds to an even number. Independently of the synthesis process variations, the number of ALMs in the Key-Bypass trojan increases 7% with respect to the clean implementation, while in ASIC, the number of combinational cells increases only 1%.

## 2.6 Related Attacks and Mitigation Techniques

### *2.6.1 Other Possible Attacks*

One alternative to reduce the execution time of the proposed attack is to use a differential-linear attack with an initialization phase of four rounds of permutation as shown in [13], where it is demonstrated that differential-linear attack on Ascon has a time complexity of $2^{18}$ which means that approximately 262.144 encryptions are needed to recover the key. Using the average execution time obtained for a single encryption in the cube attack, the differential attack would take approximately 2 s to recover a key using a similar experimental setup.

### 2.6.2 Trojan Detection

Due to its low overhead, the designed trojan for Round-Reduction does not alter significantly the power consumption of the original circuit, therefore, detecting this trojan by the analysis of its side-channel effects would be challenging, especially if the trojan is not triggered during testing. Some detection techniques that evaluate changes of path delays could be effective at detecting small trojans by incorporating Built-in Self-Test (BIST) or PUF-based circuits; however, since the Round-Reduction Trojan creates a path that does not exist in the original circuit (from *bdi_type* to *rndcmp*), this path might not be tested and the trojan could pass undetected.

Pre-silicon techniques like UCI [18] can be used to detect the designed trojans during verification of the circuit because since normally, only functional input combinations are applied to the circuit, the AND gate of the trojan trigger would not be activated and would be flagged as a potential malicious circuit. Similar techniques like VeriTrust [28] and FANCI [19] can also be effective at detecting the trojans because they can detect unused input combinations.

To detect the trojans using post-silicon detection methods based on functional or structural testing, it is important to determine the effect of unused combinations of inputs in the behavior of the circuit and used them during testing to check if there is any deviation from that expected behavior.

Finally, runtime techniques can be used to detect the Round-Reduction Trojan once it is deployed in the field. For example, a runtime security monitor can keep track of the number of clock cycles that it takes to encrypt a message with a known number of data blocks so that any alteration in the number of rounds would result in an unexpected number of clock cycles and the monitor can block the operation.

### 2.6.3 Design Recommendations

From our experience designing Trojans for the hardware implementation of Ascon, we present three design recommendations that can improve the security of cryptographic systems against hardware Trojans.

1. Hardware implementation style

   Iterative implementations of cryptographic algorithms are prone to modifications of the number of rounds by changing the reference value of the round counter. On the other hand, in pipeline and unrolled implementations, different rounds are processed using different hardware resources; therefore, modifying the number of rounds requires major changes in the circuit which makes the insertion of the trojan more challenging and/or easier to detect.
2. Key management

   Another way of increasing the security of a cryptographic system is by erasing the key register after finishing an encryption operation, forcing the key to be loaded again before performing a subsequent encryption. This prevents that an application can encrypt with the same key used by another application, which thwarts attacks like the one presented in this work that relies on the permanent residence of the key inside the cipher. This, however, can reduce the performance of the system since every new encryption, even if it uses the same key, requires loading the key into the cipher. This can be mitigated if the key is deleted only during context switching by the operating system.

3. State machine implementation

   One-hot encoding of state machines is preferred over binary encoding. Binary encoding can lead to unused states which can be exploited to insert malicious operations in the circuit as shown in this work. On the other hand, in one-hot encoding, each state is implemented using one register, avoiding unused states; therefore, inserting a malicious state would always require adding more registers which would make the insertion of the trojan more difficult and easier to detect.

## 2.7 Conclusions

In this work we presented a combined attack consisting on the insertion of a hardware Trojan and a cube attack to unveil the key of a hardware implementation of the Ascon algorithm for authenticated encryption. The Trojan used for the combined attack requires minimal modifications to the original circuit which makes its detection very challenging, especially using post-silicon techniques. We also presented a Trojan which bypasses the key directly to the output of the circuit requiring 7% more logic elements than the original circuit in a FPGA implementation. These results show that the combination of a hardware Trojan and other cryptanalysis techniques like the cube attack can be more effective in terms of detection avoidance than an attack based solely on the insertion of a trojan that exposes the key directly.

## References


1. M. Bozdal, M. Randa, M. Samie, I. Jennions, Hardware Trojan enabled denial of service attack on CAN bus. Proc. Manuf. **16**, 47–52 (2018)
2. G.T. Becker, F. Regazzoni, C. Paar, W.P. Burleson, Stealthy dopant-level hardware Trojans, in *Cryptographic Hardware and Embedded Systems—CHES 2013*, Berlin, Heidelberg, 2013, pp. 197–214
3. C. Krieg, C. Wolf, A. Jantsch, Malicious LUT: a stealthy FPGA Trojan injected and triggered by the design flow, in *Proceedings of the 35th International Conference on Computer-Aided Design—ICCAD'16*, 2016, pp. 1–8
4. X. Wang, T. Mal-Sarkar, A. Krishna, S. Narasimhan, S. Bhunia, Software exploitable hardware Trojans in embedded processor, in *Proceedings of the IEEE International Symposium on Defect and Fault Tolerance in VLSI Systems*, 2012, pp. 55–58

5. Y. Jin, N. Kupp, Y. Makris, Experiences in hardware trojan design and implementation, in *2009 IEEE International Workshop on Hardware-Oriented Security and Trust, HOST 2009*, 2009, pp. 50–57
6. S. Bhasin, J.-L. Danger, S. Guilley, X.T. Ngo, L. Sauvage, Hardware Trojan horses in cryptographic IP cores, in *2013 Workshop on Fault Diagnosis and Tolerance in Cryptography*, (IEEE, Piscataway, NJ, 2013), pp. 15–29
7. M. Yoshimura, A. Ogita, T. Hosokawa, A smart Trojan circuit and smart attack method in AES encryption circuits, in *2013 IEEE International Symposium on Defect and Fault Tolerance in VLSI and Nanotechnology Systems (DFTS)*, (IEEE, Piscataway, NJ, 2013), pp. 278–283
8. M. Kanda, Practical security evaluation against differential and linear cryptanalyses for feistel ciphers with SPN round function, Berlin, Heidelberg, 2001, pp. 324–338
9. J. Guo, J. Jean, N. Mouha, I. Nikolic, More rounds, less security?, in *IACR Cryptology ePrint Archive*, vol. 2015, 2015, p. 484
10. Y. Dodis, J. Katz, J.P. Steinberger, A. Thiruvengadam, Z. Zhang, Provable security of substitution-permutation networks, in *IACR Cryptology ePrint Archive*, vol. 2017, 2017, p. 16
11. I. Dinur, A. Shamir, Cube attacks on tweakable black boxp polynomials, in *Advances in Cryptology—EUROCRYPT 2009*, ed. by A. Joux. Lecture Notes in Computer Science, vol. 5479, 2009, pp. 278–299
12. C. Dobraunig, M. Eichlseder, F. Mendel, M. Schläffer, Ascon v1.2. Submission to the CAESAR competition, 2016, pp. 302–317. https://competitions.cr.yp.to/round3/asconv12.pdf.
13. C. Dobraunig, M. Eichlseder, F. Mendel, M. Schläffer, Cryptanalysis of Ascon, 2015, pp. 371–387
14. K. Xiao, D. Forte, Y. Jin, R. Karri, S. Bhunia, M. Tehranipoor, Hardware Trojans: lessons learned after one decade of research. ACM Trans. Des. Autom. Electron. Syst. **22**, 1–23 (2016)
15. X.T. Ngo, S. Guilley, S. Bhasin, J.-L. Danger, Z. Najm, Encoding the state of integrated circuits: a proactive and reactive protection against hardware Trojans horses, in *Proceedings of the 9th Workshop on Embedded Systems Security*, 2014, pp. 7:1–7:10
16. K. Xiao, D. Forte, M. Tehranipoor, A novel built-in self-authentication technique to prevent inserting hardware trojans. IEEE Trans. Comput. Aided Des. Integr. Circuits Syst. **33**, 1778–1791 (2014)
17. S.T.C. Konigsmark, D. Chen, M.D.F. Wong, Information dispersion for Trojan defense through high-level synthesis, 2018
18. M. Hicks, M. Finnicum, S.T. King, M.M.K. Martin, J.M. Smith, Overcoming an untrusted computing base: detecting and removing malicious hardware automatically, in *2010 IEEE Symposium on Security and Privacy*, (IEEE, Piscataway, NJ, 2010), pp. 159–172
19. A. Waksman, M. Suozzo, S. Sethumadhavan, FANCI: identification of stealthy malicious logic using Boolean functional analysis, in *CCS 2013*, 2013, pp. 697–708
20. M. Rathmair, F. Schupfer, C. Krieg, Applied formal methods for hardware Trojan detection, in *2014 IEEE International Symposium on Circuits and Systems (ISCAS)*, (IEEE, Piscataway, NJ, 2014), pp. 169–172
21. J. Rajendran, A.M. Dhandayuthapany, V. Vedula, R. Karri, Formal security verification of third party intellectual property cores for information leakage, in *2016 29th International Conference on VLSI Design and 2016 15th International Conference on Embedded Systems (VLSID)*, (IEEE, Piscataway, NJ, 2016), pp. 547–552
22. A. Amelian, S.E. Borujeni, A side-channel analysis for hardware Trojan detection based on path delay measurement. J. Circuits Syst. Comput. **27**, 1850138 (2018)
23. X. Wang, H. Salmani, M. Tehranipoor, J. Plusquellic, Hardware Trojan detection and isolation using current integration and localized current analysis, in *2008 IEEE International Symposium on Defect and Fault Tolerance of VLSI Systems*, (IEEE, Piscataway, NJ, 2008), pp. 87–95
24. (2014, 16/05/2019). *CAESAR call for submissions, final (2014.01.27)*. Available: https://competitions.cr.yp.to/caesar-call.html

25. Z. Li, X. Dong, X. Wang, Conditional cube attack on round-reduced ASCON. IACR Trans. Symmetric Cryptol. **2017**, 175–202 (2017)
26. E. Homsirikamol, Ascon hardware, 17 Oct 2016. Available: https://github.com/IAIK/ascon_hardware/tree/master/caesar_hardware_api_v_1_0_3/ASCON_ASCON
27. A. Cui, X. Qian, G. Qu, H. Li, A new active IC metering technique based on locking scan cells, in *2017 IEEE 26th Asian Test Symposium (ATS)*, 2017, pp. 40–45
28. J. Zhang, Q. Xu, Z. Sun, F. Yuan, L. Wei, VeriTrust, in *2013 50th ACM/EDAC/IEEE Design Automation Conference (DAC)*, 2013, p. 1

# Chapter 3
# Anti-counterfeiting Techniques for Resources-Constrained Devices

**Yildiran Yilmaz, Viet-Hoa Do, and Basel Halak**

## 3.1 Introduction

The perceived advantages of RFID systems in reducing the risk of counterfeiting and forgery can be seriously weakened by emerging security attacks such as tag cloning on these systems. Counterfeit products are particularly problematic in safety-critical applications, for example, in pharmaceutical products where a counterfeit drug causes a loss of life [1]. Since RFID systems aim to authenticate a tagged item, RFID tags are used to prevent counterfeiting of products such as medicines from some health industry brands or government documents [2]. In the instance of pharmaceutical products, to prevent the counterfeiting of a drug, the manufacturer aims to authenticate the identity of the drug by placing an RFID tag on it with specific and reference information. When a drug passes a reader, at the point of sale, the reader checks whether the required reference information is available and valid on the tag. This check is accomplished by a security protocol between the tag and the reader. If the necessary information is authentic, the drug is stated to be genuine.

Regarding all possible security threats, RFID has improvements over traditional anti-counterfeiting mechanisms, e.g. holograms and barcodes [2]. The advantages of RFID systems over these currently used identification systems can be summarised as follows [2]. RFID systems have large operating and communication range, read and write capability of the transponder memory, data encryption/authentication capability, and provides hands-free operation. Hereby RFID-tagged products can be accurately identified without physical and visual contact. Therefore, RFID technology reduces theft and provides tracking and dynamic pricing of products without affecting supply chain efficiency [3]. As RFID technology is needed primarily

Y. Yilmaz · V.-H. Do · B. Halak (✉)
University of Southampton, Southampton, UK
e-mail: yildiran.yilmaz@erdogan.edu.tr; doviethoa@doviethoa.com; basel.halak@soton.ac.uk



B. Halak (ed.), *Hardware Supply Chain Security*,
https://doi.org/10.1007/978-3-030-62707-2_3

in applications such as fraud protection, secure access and anti-counterfeiting, authentication protocols are the cornerstone of ensuring both the data privacy and integrity of RFID systems.

In order to resist counterfeiting, an item is usually authenticated remotely or semi remotely. In authentication, the party that verifies the tagged product can obtain tag information. However, since RFID systems are ubiquitous technology, the attacker can tamper with tag information and deceive genuine readers. To trust an RFID authentication system, it is essential to maintain the confidentiality and integrity of tag data. Therefore, security protocols should consider tag data privacy and integrity (anti-counterfeiting). Admittedly, while conventional protocol solutions based on traditional crypto algorithms have proven to be resistant to cryptanalysis, performing such protocols require a great deal of computation and communication resources which might not be possible to implement in resource-constrained devices, e.g. RFID tags [4].

While a number of lightweight authentication protocols have been proposed in [5–9], none of them offers a complete security solution in the context of the key three qualities of lightweight mutual authentication, availability and tag unclonability. Therefore, none of them is truly secure solution in providing fraud protection, secure access and anti-counterfeiting. For example, none of the schemes proposed in [5, 6] or [7] protect against physical cloning attacks. The WIPR protocol proposed by Arbit et al. [5] uses the shared public key preinstalled in the tag memory and is vulnerable to tag cloning attack and does not provide mutual authentication. On the other hand, Fu et al. [6] employ a symmetric encryption to guarantee data confidentiality and integrity, but managing and storing the secret key is a major weakness and the system could be vulnerable if this key is compromised [10]. There are other approaches which rely on the use of PUF technology, which enhances the resilience of RFID systems against tag cloning attacks [8, 9]. Gope et al. [8] proposed a mutual authentication protocol using a Hash function and a PUF. Although it is claimed in [8] that the protocol satisfies all common security requirements including availability, this protocol cannot fully resolve the availability issue posed by a desynchronisation attack. Chatterjee et al. [9] proposed a mutual authentication protocol based on PUF and ECC. However, as stated in [11], Chatterjee's protocol still cannot protect the tags' anonymity and their availability, and forward security has not been proved.

As such, none of the aforementioned protocols can simultaneously provide the three qualities of lightweight mutual authentication, availability and tag unclonability. To this end, this chapter reports the following contributions:

1. A new lightweight mutual authentication protocol that combines Rabin public-key encryption with PUF technology in order to introduce the ability for the reader to securely transmit the different public keys to the tag in each transaction, and to prevent tag cloning. The security of Rabin cryptosystem [12] relies on the difficulty of large integer factorisation similarly to the RSA scheme. The prominent advantage of Rabin is the much simpler encrypting operation (modular squaring) than RSA and ECC [13]. Therefore, the encryption process

is performed on the tag side, while the decryption process, which requires further processing, is performed on the reader side.
2. A detailed hardware design of the security system of the RFID tag.
3. A systematic evaluation of the proposed protocol and tag hardware design against security attacks, namely eavesdropping, tracking, reader impersonation, desynchronisation, replay and tag cloning.

*The context of the experiment*: The experiment is carried out to estimate the area-related cost of the proposed protocol. The estimation result is then evaluated by comparing it with other existing protocols.

*Assumptions*: This chapter makes the following assumptions:

- The means of communication is a radio wave.
- The RFID tags can be deployed in a publicly accessible environment.
- All the tags deployed are resource-constrained in terms of area and computation.
- The reader is comparatively a resource-rich device in terms of area, energy, memory and computation ability.

## 3.2 Chapter Overview

The organisation of this chapter is as follows. In Sect. 3.3, the background information on Rabin cryptosystem and related work are provided. Sections 3.4 and 3.5 present the proposed protocol and the security analysis of the system. Subsequently, the tag hardware design is described in Sect. 3.6, whereas Sect. 3.7 evaluates the system performance and design cost. Finally, Sect. 3.8 concludes the chapter.

## 3.3 Background

This section describes and discusses the methods used in this chapter. The first part describes the Rabin encryption and decryption and evaluates the resource usage of its implementations. The second part discusses the related work.

### *3.3.1 Rabin Cryptosystem*

The Rabin scheme, proposed in [12], is an asymmetric cryptosystem that relies on the difficulty of the factorisation of large integer numbers. It is equivalent to RSA in term of security and computational complexity [12]. However, it is much simpler to perform the Rabin encryption process compared to RSA, while the Rabin decryption requires more computation and resources than RSA [13].

- A Rabin cryptosystem's private key consists of two large distinct primes $p$ and $q$. In order to simplify the decryption, the following condition is used:

$$p \equiv q \equiv 3 \pmod 4 \tag{3.1}$$

- The public key referred to as $n$ is the product of the private key pair:

$$n = p \times q \tag{3.2}$$

- Rabin encryption is simply a modular squaring operation, which is much less complex compared to RSA or ECC at the same security level. This calculation is carried out to generate a cypher text ($C$) from a plain text ($M \in 0, 1, \ldots, n-1$), as follows:

$$C = M^2 \pmod n \tag{3.3}$$

- Rabin decryption finds the square root of the cypher text $C$ modulo $n$ using the Chinese remainder theorem [5]. Under the condition (3.1), the calculation of the square root of $C$ modulo $p$, $q$ is as follows:

$$\begin{aligned} m_p &= \sqrt{C} = C^{\frac{p+1}{4}} \pmod p \\ m_q &= \sqrt{C} = C^{\frac{q+1}{4}} \pmod q \end{aligned} \tag{3.4}$$

- Let $y_p$, $y_q$ be the Bzout's identity of $p$ and $q$, the four square roots $m_0$, $m_1$, $m_2$, $m_3$ of $C$ modulo $n$ are then calculated as follows:

$$\begin{aligned} m_0 &= \left(y_p \times p \times m_q + y_q \times q \times m_p\right) \bmod n \\ m_1 &= n - m_0 \\ m_2 &= \left(y_p \times p \times m_q - y_q \times q \times m_p\right) \bmod n \\ m_3 &= n - m_2 \end{aligned} \tag{3.5}$$

- The plaintext $M$ will be one of the four square roots. Additional information, for example, a known part of the plaintext ($M$), is required to choose the correct one.

In order to reduce memory usage, a randomised variant of Rabin encryption avoiding the complex modular operation was proposed by Shamir [14]. The encryption in Shamir approach is as follows. If $R$ is a random number with a bit width longer than the key size, the cypher text is computed such that:

$$C = M^2 + R \times n \tag{3.6}$$

Other variants of Rabin include WIPR [5] which is an application of Rabin encryption specifically designed and optimised for lightweight passive RFID tags employing the randomised multiplication described above. In 1024-bit Rabin encryption of the WIPR design, the area-related implementation cost is 4184 GEs which is the data path area.

### 3.3.2 Related Work

PUFs are an integrated circuitries which produce different outputs when they are implemented in different devices [15]. Manufacturing process variabilities cause the variation which is unavoidable and uncontrollable in the property of circuits, and serves PUF's unclonability property. PUFs have been employed in [16, 17] for key generation and in [18, 19] for anti-counterfeiting, IP (intellectual property) protection.

As PUFs are a device-specific function and offer the unclonability property, they have been gaining attention in the RFID field for authentication purposes. Several studies, such as those reported in [8, 9], have been conducted on PUF-based authentication. The authors in [8] developed a new method for an authentication protocol using the hash function and PUF. Their approach is based on using secret data shared synchronously between the reader and the tag as essentially a root of security. Nevertheless, such protocols based on shared secrets are usually subject to desynchronisation attacks unless a solution for this vulnerability is provided. To address this, the authors in [8] suggested a solution that can provide resistance against desynchronisation attacks. The solution is based on a protocol whereby a series of challenge-response pairs called emergency CRPs stored on the reader side, and identities named unique unlinkable pseudo IDs stored on the tag side are used to overcome any possible synchronisation attack. However, this method suffers from two pitfalls. First, these CRP sets are limited and have to be updated over time. Secondly, this data set requires large amounts of storage that are not suitable for the area-constrained design of the tag.

Elsewhere, the authors in [9] developed another authentication protocol using ECC and PUF, which is the most complex protocol design so far. They used the PUF to generate public keys for ECC to provide secrecy of the public keys. Unfortunately, they fail to explain how forward secrecy is satisfied in their protocol. Nevertheless, it has been presented in [11] that this protocol does not meet the anonymity requirement. Also, the work reported in [11] explains that this protocol is not secure against impersonation, man-in-the-middle and replay attacks.

In summary, the aforementioned discussion describes the required cryptographic primitive, namely the Rabin cryptosystem for the proposed protocol, and deduces that there is a need for a highly secure and lightweight mutual authentication protocol for RFID systems. In the following section, the proposed protocol for RFID system is described in detail.

## 3.4 Proposed Protocol

In this section, the proposed solution is overviewed and the system model is described along with the threat assumptions in order to identify the security requirements that the proposed protocol meets. Section 3.5 will later explain how these requirements have been proved by the security analysis. Furthermore, this Sect. 3.4 subsequently describes the proposed authentication protocol in detail and discusses the recommendations for protocol security level.

### *3.4.1 Proposed Solution*

The use of chip area in RFID systems should be considered when applying security requirements as the RFID tags are resource-constrained devices. While security requirements [10] such as mutual authentication, forward security, confidentiality, availability, anonymity and tag unclonability are significant for the security of protocols, these requirements must be met in RFID systems by taking into account the chip area. Accordingly, our proposed solution offers a new security protocol utilising PUF technology and Rabin encryption unitedly that meets the security requirements mentioned above, considering area consumption.

In cryptography term, Rabin scheme [12] is established as a way to encrypt and decrypt message. In our solution, Rabin is implemented so that secret data is encrypted by the tag and decrypted by the reader because Rabin encryption requires much less computation than other asymmetric encryption algorithms (e.g. RSA and ECC) [13]. Another cryptographic system PUF is used in our solution to produce tag-specific secrets to be enciphered by Rabin. As a result, PUF and Rabin cryptographic principles are used to provide data confidentiality, forward security, anonymity and resist attacks; modification, eavesdropping and tag cloning.

### *3.4.2 System Model*

In the system model of this chapter, shown in Fig. 3.1, the RFID tag is mutually authenticated with the reader. The tag also passes its data to the reader in encrypted form using public-key encryption over the insecure channel. The tags are resource-constrained parties, but the readers and the database behind the reader are considered not to have a major restriction in terms of design cost and resource consumption. Moreover, this chapter focuses on the tag and its communication with the reader over the insecure channel; therefore, the tag considers other entities of the system such as the central server, database as part of the reader.

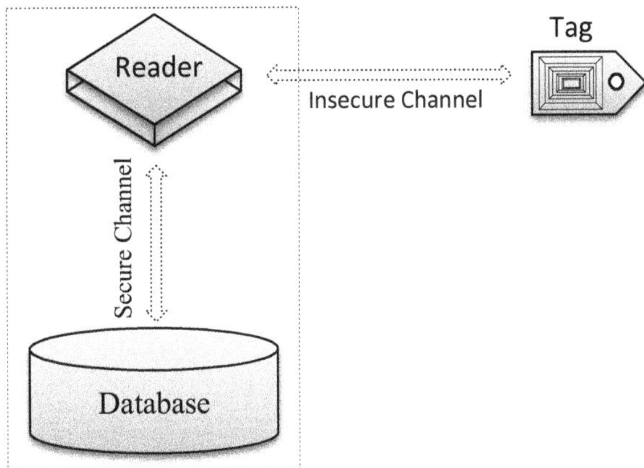


**Fig. 3.1** RFID system model

### 3.4.3 Attacker Model and Threat Assumptions

- Server database and its connection with the reader shown in Fig. 3.1 are considered to be trusted. The database is considered as the resource-rich party so that it can work in a secure manner. The adversary can impersonate the reader by using a malicious reader, but it cannot access the tag database in the trusted system.
- In attacker model, the communication shown in Fig. 3.1 between the tags and readers can be accessed and tampered by the adversary, including listening to all messages, jamming the air interface and modifying any communication.
- The adversary has full knowledge about the protocol, as well as all the security primitives and algorithms used by the tag. It is assumed that the PUF is designed with tamper-resistant because the effort to probe it including the physical attacks to the internal PUF changes the PUF itself or destroys it. However, the adversary can apply a random input to the tag interface to monitor output.

### 3.4.4 Main Design Aims

The main security properties that the proposed protocol aims to meet are defined below. Note that Rabin cryptosystem described in Sect. 3.3.1 in the proposed protocol is developed based on the WIPR design [5].

- *Mutual authentication*: the proposed protocol should allow both communicating parties (the reader and tag) to authenticate each other. In the WIPR design described in [5], there is no mutual authentication property.
- *Public-key transmission*: the proposed protocol should allow parties to use different public and private keys in each authentication cycle. To achieve this, the proposed protocol should allow the reader to transmit the public key to the

Table 3.1 Notations used in this chapter

| Notation | Description |
|---|---|
| $a\|b$ | Concatenation of $a$ and $b$ |
| $H(x)$ | Hash function |
| PUF($x$) | PUF response of challenge $x$ |
| $p, q$ | Rabin private keys |
| $n, s_n$ | Rabin public key and its signature |
| $uid$ | Unique ID |
| $tid, s_{tid}$ | Temporary ID and its signature |
| $R_t$ | Random number generated by the tag |
| $H_r, H_t$ | Message hash |
| $C$ | Encrypted message |
| $M$ | Plain message |
| $k$ | Transaction number |

tag, and this transmission should guarantee the integrity of the public key using a secure method. Besides, the tag should verify the authenticity of the public key.
- *Unclonability*: the tag design and the tag unique ID should be able to resist against cloning attacks even if an attacker knows the tag design details.

### 3.4.5 Proposed Mutual Authentication Protocol

The proposed protocol combines PUF technology with an asymmetric cryptosystem, namely the Rabin cryptosystem described in Sect. 3.3.1, to produce a new protocol for RFID systems. The choice of the Rabin scheme was mainly driven by cost implications, as mentioned later in Sect. 3.1. However, the proposed protocol can be used with any asymmetric cryptosystem. The notations used in the protocol description are summarised in Table 3.1.

The protocol consists of three phases, namely registration, identification and verification. In the registration phase, it is assumed that a random seed *tid* for the temporary ID (TID) is stored on the tag. The system stores the following authentication information of the first $K$ transactions in the database for future use:

- The tag unique identifier *uid*
- Authentication information of the first $K$ transactions:
  - Transaction number $k = 1, 2, \ldots$
  - The TID and its signature $s_{tid} = \text{PUF}(tid)$
  - The private key $(p, q)$ and the signature $s_n = \text{PUF}(H(n\|tid))$ of the corresponding public key $n = p \times q$

The proposed authentication protocol is depicted in Fig. 3.2.

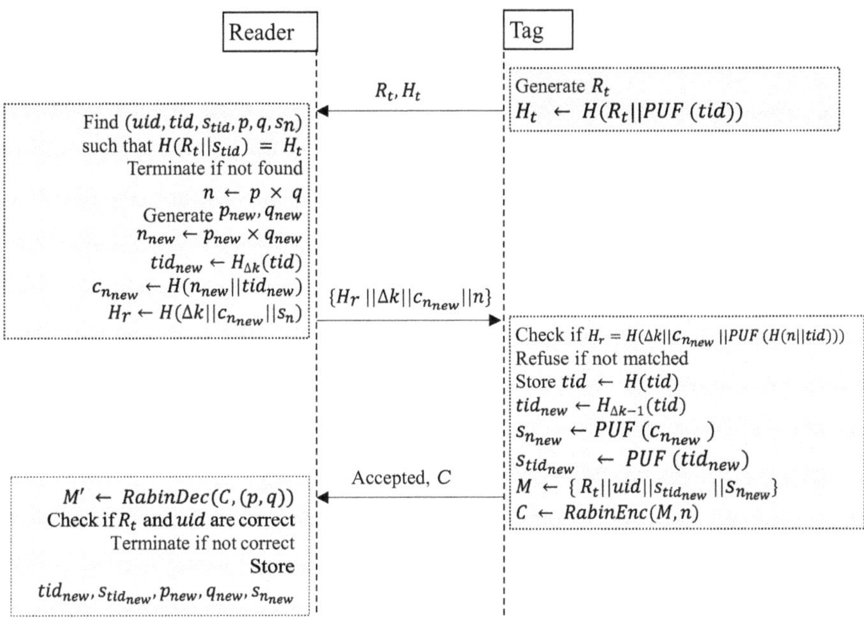


**Fig. 3.2** The proposed mutual authentication protocol

The following steps are carried out to perform the proposed mutual authentication between the reader and the tag.

### 3.4.5.1 Phase 1: Identification

The reader first identifies the tag using TID in the following two steps:

1. The tag generates a random number $R_t$. The hash value $H_t$ of the random number $R_t$ and the signature $s_{tid} = \text{PUF}(tid)$ of the current *tid* is transmitted to the reader:

$$H_t = H\left(R_t \middle\| \text{PUF}(tid)\right) \tag{3.7}$$

2. The reader searches the database such that $H(R_t \| s_{tid}) = H_t$ ($R_t$, $H_t$ are from the tag, $s_{tid}$ is from each database entry). If the query cannot be found, the reader terminates the authentication process. This identification is carried out based on the PUF response $s_{tid}$; in this way, the reader distinguishes the PUFs used by the tag. If the query is found, the reader then uses the tag data stored in the database to perform authentication in Phase 2. The tag data stored in the database contains the following information:

- The unique identifier *uid*
- The transaction number $k$
- The current TID *tid* and its signature $s_{tid}$
- The current private key $(p, q)$ and the signature $s_n$ of the corresponding public key

#### 3.4.5.2 Phase 2: Authentication

Both party (reader and tag) authenticate each other exploiting the shared secrets in the following three steps:

1. Let $k_{\text{oldest}}$ and $k_{\text{newest}}$ be the oldest and the newest transaction information of the same tag in the database. By comparing the transaction numbers $k$ and $k_{\text{oldest}}$, the reader knows the number of transactions the tag has drifted by from the previous successful one is $(k - k_{\text{oldest}})$. Depending on the system-specific security policy, it could then decide whether to proceed with the current authentication or perform other emergency actions such as activating the alarm. If the security policy requires the system to be non-tolerant to out-of-sync situations, the proposed protocol can detect an out-of-sync state by checking whether $k > k_{\text{oldest}}$ or not. The tag is supposed to be in the transaction, which $k = k_{\text{oldest}}$, if there is no desynchronisation attack. But if an attack desynchronised the tag and the reader, the tag may drift to transaction $k > k_{\text{oldest}}$. If the authentication process continues, the reader prepares to ask the tag for the authentication information of the $\Delta k$ _ next transaction, where $\Delta k = k_{\text{newest}} - k + 1$. The reader expects the tag to send back the authentication information for the $(k + \Delta k)$th $= (k_{\text{newest}} + 1)$th transaction.

   The reader computes the public key $n$ from the private key $(p, q)$ for this transaction:

   $$n = p \times q \tag{3.8}$$

   The reader then generates a new Rabin private key $(p_{\text{new}}, q_{\text{new}})$ and computes the corresponding public key $n_{\text{new}}$ in the following formula. The $p_{\text{new}}$ and $q_{\text{new}}$ are two large distinct primes that satisfy the condition in Eq. (3.1).

   $$n_{\text{new}} = p_{\text{new}} \times q_{\text{new}} \tag{3.9}$$

   The TID and PUF challenge for the $\Delta k$-next transaction is computed as follows:

   $$tid_{\text{new}} \leftarrow H_{\Delta k}(tid) \tag{3.10}$$

   $$c_{n_{\text{new}}} \leftarrow H\left(n_{\text{new}} \middle\| tid_{\text{new}}\right) \tag{3.11}$$

Note that $H_{\Delta k}(tid) = H(H \ldots (tid))(\Delta k$ times $H)$.

The reader computes the message signature $H_r$ and transmits $n$, $c_{n_{new}}$ and $H_r$ to the tag:

$$H_r \leftarrow H\left(\Delta k \,\|c_{n_{new}}\|\, s_n\right) \tag{3.12}$$

2. The tag checks if the message signature $H_r$ from the reader and the value computed by the tag itself match one another. If they do, the reader is authentic as it provides the correct $s_n$, which is secret, and the message itself is intact. Otherwise, the tag transmits a refusal message to the reader and terminates the authentication process.

$$H_r = H\left(\Delta k \,\|c_{n_{new}}\|\, \mathrm{PUF}\left(H\left(n\middle\|tid\right)\right)\right) \tag{3.13}$$

Having authenticated the reader, the tag then stores the next TID to the *tid* memory and computes the secret information for the $\Delta k$-next transaction. The tag computes the secret information of the next transaction as follows:

$$\begin{aligned} tid &\leftarrow H(tid) \\ tid_{new} &\leftarrow H_{\Delta k-1}(tid) \\ s_{n_{new}} &\leftarrow \mathrm{PUF}\left(c_{n_{new}}\right) \\ s_{tid_{new}} &\leftarrow \mathrm{PUF}\left(tid_{new}\right) \end{aligned} \tag{3.14}$$

The tag encrypts and then transmits its *uid* along with $s_{tid_{new}}$, $s_{n_{new}}$ to the reader using Rabin encryption with the public key $n$. The randomised variant of Rabin encryption in Eq. (3.6) is used. When the current authentication is successfully completed, $n_{new}$ is going to be used in the next authentication process, but $n$, $p$ and $q$ are going to be discarded.

$$\begin{aligned} M &\leftarrow \left\{R_t \,\|uid\|\, s_{tid_{new}} \middle\| s_{n_{new}}\right\} \\ C &\leftarrow \mathrm{RabinEnc}(M, n) \end{aligned} \tag{3.15}$$

3. The reader decrypts the message $C$ with the private key $(p, q)$ using Eq. (3.5). $R_t$ is the known part of the plaintext which can be used to find the correct square root.

$$M' \leftarrow \mathrm{RabinDec}(C, (p, q)) \tag{3.16}$$

If *uid* sent by the tag is the same as the value stored in the reader, the tag is authentic and fully identified. Otherwise, the authentication process is terminated. If all the authentication steps are successfully completed, then the reader stores the secret information $tid_{new}$, $s_{tid_{new}}$, $(p_{new}, q_{new})$ and $s_{n_{new}}$ in the database for future use and removes the old secret information *tid*, $s_{tid}$, $(p, q)$

and $s_n$. This happens when the third flight is not disrupted, and $C$ are decrypted correctly ($R_t$ and $uid$ are correct). But, if the third flight is disrupted after tag computes $C$, then the reader will not store new secret information and keeps the current secret information in database.

### 3.4.6 Data Retrieval Protocol

By using Rabin encryption, it is possible to securely retrieve data from the tag. One of the benefits of this is that it allows the reader to constantly maintain the authentication information of the tag for the next $K$ transactions and recover from a desynchronisation attack.

The secret request protocol is depicted in Fig. 3.3. Each step in the secret request protocol is mostly the same as the authentication protocol.

- The reader generates a new private key $p_{\text{new}}, q_{\text{new}}$, the TID $tid_{\text{new}}$ of the $\Delta k$-next transaction, the PUF challenge$c_{n_{\text{new}}}$ and the message checksum $H_r$ and then sends them to the tag.
- The tag checks if the checksum $H_r$ is correct. If so, it computes the secret authentication information $s_{n_{\text{new}}}$ and $s_{tid_{\text{new}}}$ and encrypts them using the current transaction public key $n$ and subsequently sends them in encrypted form back to the reader.
- The reader decrypts the message using the current transaction private key $(p, q)$ and the known part of the plaintext ($uid \oplus H_r$). If the decryption succeeds, it stores the new secret information $tid_{\text{new}}, s_{tid_{\text{new}}}, (p_{\text{new}}, q_{\text{new}})$ and $s_{n_{\text{new}}}$ in the database for future use.

### 3.4.7 Recommendations for Security Level

Confidential data and the security algorithms involved in the protocol ought to be at the same security level to the Rabin public key size recommended by NIST [20]. To give an example, in a 1024-bit Rabin encryption, the PUF and Hash function should have at least 80-bit outputs. As will be analysed in Sect. 3.5.4, the proposed protocol prevents an attacker from accessing any CRP generated by PUF since it hides CRP accessibility by encryption, therefore using a lightweight PUF, e.g. Arbiter-PUF is possible [21]. The hash function, termed $H$, is employed to generate the TID, PUF challenges and message checksum. The extra protection of the PUF allows the hash function to be weaker but still maintains the sufficient security level.

Thus far, the details on the description of the proposed mutual authentication protocol for RFID system are provided in this section. The next section analyses in detail the security of the proposed protocol outlined above.

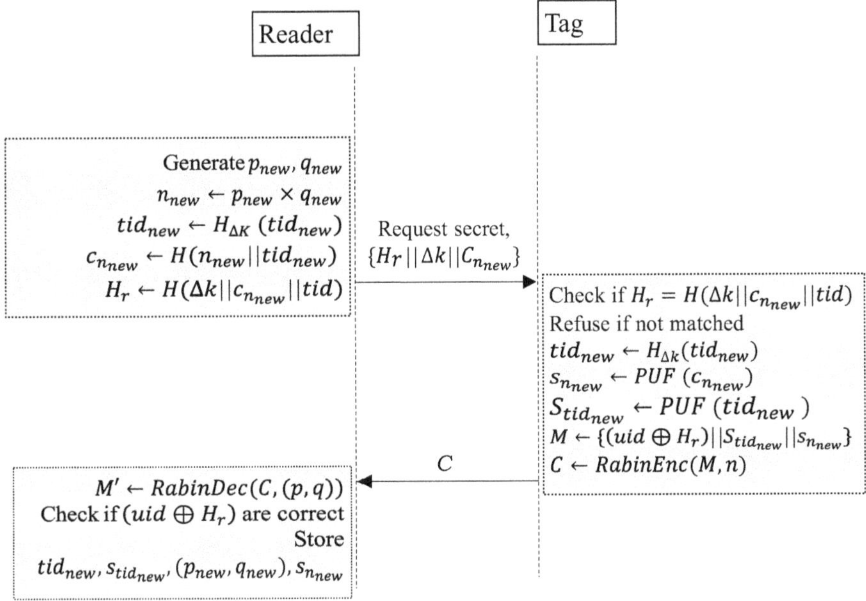


**Fig. 3.3** Secret request protocol

## 3.5 Security Analysis

The tag and reader interaction over the insecure environment may encounter various attacks [22] as listed below.

- Man in the middle
- Eavesdropping
- Tracking
- Reader impersonation
- Desynchronisation
- Replay
- Tag cloning

Therefore, this section describes the security analysis of our protocol to validate the proposed solution. It begins with a description of the security requirements and subsequently defines the interpretation of the proposed protocol with Scyther and explains the verification of security requirements. Finally, it compares the proposed protocol with other related work on security protocols.

**Table 3.2** Security analysis of the proposed protocol and other authentication protocols

| Protocols | [5] | [8] | [9] | Proposed |
|---|---|---|---|---|
| *Security requirements* | | | | |
| Mutual authentication | ○ | ● | ● | ● |
| Confidentiality | ● | ● | ○ | ● |
| Anonymity | ● | ● | ○ | ● |
| Availability | ● | ◐[a] | – | ● |
| Forward security | ● | ● | – | ● |
| *Types of attack* | | | | |
| Eavesdropping attack | ● | ● | ● | ● |
| Replay attack | ● | – | ● | ● |
| Tracking attack | ● | ○ | ○ | ● |
| Desynchronisation attack | ● | ◐[a] | ● | ● |
| Reader impersonation | ● | ● | – | ● |
| Tag cloning | ○ | ● | ● | ● |
| Man in the middle | ● | ● | ○ | ● |

●: Fully satisfied, ◐: Partially satisfied, ○: Unsatisfied, –: Not provided

[a]Mitigated using emergency CRPs

## 3.5.1 *Security Requirements*

Several security requirements for the communication between RFID tags and readers must be satisfied as mentioned in [10]. Each security requirements from Sects. 3.5.1.1–3.5.1.5 will be defined in this subsection and analysed systematically with the Scyther tool [23] in Sect. 3.5.2. The tag unclonability requirement will be analysed in Sect. 3.5.4. The summary of the proposed protocol security analysis is presented in Table 3.2.

### 3.5.1.1 Mutual Authentication

In the context of mutual authentication, the tag and the reader must be able to verify each other's authenticity before any sensitive data is exchanged. The proposed protocol satisfies mutual authentication in the following manner.

The tag first verifies the reader's authenticity by comparing the message signature $H_r$ sent by the reader and the signature computed by the tag itself (see Eq. 3.12). Only the tag can generate the correct signature of the public key $S_n$ because of the PUF, and the tag only sends $S_n$ to the trusted reader, therefore only the trusted reader can generate the correct message signature $H_r$. The reader then verifies the tag's authenticity by looking up the TID and the unique identity *uid* sent by the tag. Only an authentic tag which has the matched TID and PUF CRPs can produce the correct values.

#### 3.5.1.2 Confidentiality

To ensure confidentiality, all secret and sensitive information must be transmitted securely. The proposed protocol satisfies confidentiality since all secret information of the tag including the unique identifier *uid*, the signature of the next-TID $s_{tid_{\mathrm{new}}}$ and the public key $s_{n_{\mathrm{new}}}$ are always encrypted before transmitting them over the wireless channel.

#### 3.5.1.3 Anonymity

To ensure anonymity, it is required that the tag cannot be traced by the adversary. The proposed protocol satisfies anonymity since the adversary cannot know the true UID of the tag because it is always encrypted before transmission. The TID is protected and randomised in each transaction by PUF; therefore, there is no correlation between the TID and its true identity.

Other unprotected data such as $R_{\mathrm{t}}$, $c_{n_{\mathrm{new}}}$ and $H_{\mathrm{r}}$ are random by nature, and there is no correlation between these values in one transaction and in any other transaction, even when the tag internal state *tid* is not changed.

#### 3.5.1.4 Availability

In an RFID system, attacks on availability could be replay or desynchronisation attacks as discussed in [24]. Therefore, a security protocol needs to provide resistant to such attacks to maintain availability under desired levels inquired by RFID application. The proposed protocol satisfies availability because the protocol can resist replay and desynchronisation attacks. To prevent a replay attack, all the messages are randomised for each tag and each authentication cycles in the proposed protocol. To prevent desynchronisation attack, both the tag and the reader remain synchronised and ready to communicate in the following way. By storing authentication information for $K$ consecutive transactions, the reader can detect if the tag and the reader are desynchronised by comparing $k$ with $k_{\mathrm{oldest}}$ (Note that in the case of a desynchronisation attack, $k > k_{\mathrm{oldest}}$). After this attack is detected, the reader may query to run the secret request protocol demonstrated in Fig. 3.3 to retrieve authentic secret information from the tag. Furthermore, after any successful authentication, the reader can ask the tag for more authentication information, refills the safe margin of $K$ consecutive transactions and recovers from the attack completely.

#### 3.5.1.5 Forward Security

To guarantee forward security, it is required that the prior communications are secure even if secret info is leaked in the present. Our protocol satisfies forward security in the following manner. The tag random-TID *tid* and its true identifier *uid* are two important secret information of the tag. However, since the TID is protected by the PUF before it is transmitted, it is not possible for the adversary to identify the tag even when the TID random seed *tid* has been compromised. The unique ID (*uid*) is always transmitted securely using Rabin encryption. Even when the *uid* is compromised, it is not possible to identify the tag from the encrypted message $C$ because the plaintext $M$ is padded with random numbers generated by both the reader and the tag and encrypted using the random public key provided by the reader.

#### 3.5.1.6 Tag Unclonability

To ensure tag unclonability, it is essential to secure the data on the tag and the data transmitted wirelessly on the channel. Tag cloning can be done by physically tampering with the tag and printing properly formatted data to the blank RFID tag or by obtaining the data transmitted in the channel. In the proposed protocol, both the PUF cryptographic primitive, which secures data on the tag, and the PUF response-hiding scheme, which secures the transferred data on the channel, provide tag unclonability. Rabin encryption in the PUF response-hiding scheme shown in Fig. 3.5 secures the transferred data by encrypting all the sensitive data *uid*, $s_{n_{\text{new}}}$ and $s_{tid_{\text{new}}}$ which both $s_{n_{\text{new}}}$ and $s_{tid_{\text{new}}}$ are PUF responses.

### *3.5.2 Analysis of the Proposed Protocol with Scyther*

As mentioned in [23], Scyther is an open-source security verification tool which can analyse authentication protocols in a systematic manner. Therefore, to validate the proposed solution, the proposed protocol was defined with Scyther language as shown in figure. The protocol includes the roles of the objects involved, public and private variables, databases, transmitted and received messages between objects and the order of these exchanged messages. Most importantly, each authentication protocol possesses security properties that must be proved in a systematic manner. In the tool, these properties are called "claim events". Unless the value of the variable is not exposed to the adversary, a *secrecy* claim of a variable can remain protected when two reliable parties are in communication with each other. The non-injective synchronisation claim, namely *Nisynch* in the Scyther language, means that all processes expected to occur in the theoretical definition of the protocol should be performed without error and interruption during the execution of the protocol. This claim shows that the received messages are sent by the sender and the sent messages are received by the receiver. The *commit* claims described in Listing 3.1 show that all communication partners have agreed on the variable values.

**Listing 3.1: Definition of the Proposed Protocol in Scyther**

```
//proposed protocol description
  const PUF: Function;
  hashfunction H1;
  secret uid: Function;
  // sk(X) denotes the long-term private key of Reader
  // pk(X) denotes the corresponding public key
  protocol RFIDprotocol(Tag, Reader)
  {
    role Tag
      {
        fresh Rt: Nonce;
        fresh tid: Nonce;
        fresh stidnew: Nonce;
        var tidnew;
        var n: Nonce;
        var snnew: Nonce;
        var cnnew: Nonce;
        var Ht;
        var Hr;

        match(Ht, H1(Rt, PUF(k(Tag, Reader), tid)));
        send_1( Tag, Reader, Rt, Ht);
        recv_2(Reader, Tag, Hr, cnnew, n);
        match(snnew, PUF(k(Tag, Reader), cnnew));
        macro tidnew = H1(tid);
        match(stidnew, PUF(k(Tag, Reader), tidnew));
        send_3(Tag, Reader,{Rt, uid, stidnew, snnew}pk(Reader));

        claim_Tag1(Tag, SKR, snnew);
        claim_Tag2(Tag, SKR, stidnew);
        claim_Tag3(Tag, Nisynch);
        claim_Tag4(Tag, Commit, Reader, n, cnnew, Hr, Ht, Rt, snnew, stidnew);
      }
    role Reader
      {
        fresh n: Nonce;
        fresh cnnew: Nonce;
        fresh Sn: Nonce;
        var Hr;
        var Ht;
        var Rt;
        var snnew: Nonce;
        var stidnew;

        recv_1(Tag, Reader, Rt, Ht);
```

```
            match(Hr, H1(cnnew, Sn));
            send_2(Reader, Tag, Hr, cnnew, n);
            recv_3(Tag, Reader,{Rt, uid, stidnew, snnew}pk(Reader));

            claim_Reader1(Reader, SKR, snnew);
            claim_Reader2(Reader, SKR, stidnew);
            claim_Reader3(Reader, Nisynch);
            claim_Reader4(Reader, Commit, Tag, n, cnnew,Sn, Hr, Ht, Rt, snnew,
stidnew);
        }
    }
```

Listing 3.1 above defines the proposed RFID authentication protocol in the Scyther programming language. There are two roles: the tag's, i.e. *role Tag*, and the reader's, i.e. *role Reader*. The local variables of each role are defined as the nonces $R_t$, $tid_{\text{new}}$, $s_{tid_{\text{new}}}$, $n$ and the challenge $c_{n_{\text{new}}}$ and the response $s_{n_{\text{new}}}$. The tag and the reader each produce a single nonce, $R_t$ and $tid_{\text{new}}$, respectively. The challenge $c_{n_{\text{new}}}$ is generated by the tag as a nonce, and the responses $s_{n_{\text{new}}}$ and $s_{tid_{\text{new}}}$ are generated by the PUF declared as a constant function, which is only known to the tag and the reader.

Each variable is either specified by declaring it inside the role as fresh or var. The fresh declaration, as well as a variable such as nonce, are randomly generated numbers which are local to the role. The values declared as var denote an assigned value upon receipt of a message. Send and receive events are used in order to send and receive messages, respectively, between roles. In communication events such as *send* and *recv*, the first parameter denotes the identity of the sender, and the second indicates the identity of the receiver. The rest of the parameters specify the content of the message containing variables. It is required that every variable in the *send* event is given a value. The values from the contents of the *send* event are also assigned to the variables in the *recv* event. The *match* event is used for pattern matching in a way that it assigns the value of the second parameter to the first parameter. Thus, via the match event, the output of the PUF function, which is the response, is computed on the secret key shared between the tag and the reader, i.e. *k*(Tag, Reader), and on the challenge $c_{n_{\text{new}}}$. The *macro* event defines abbreviations for a particular function to simplify the protocol specification, while the *claim* events define the intended security properties as previously mentioned in Table 3.2. The tool automatically verifies the security properties described in *claim* events as shown in Fig. 3.4.

### 3.5.3 *Verification of Security Requirements*

The results of the security analysis on the proposed authentication protocol using the Scyther tool showed that all involved claims are verified, as illustrated in Fig. 3.4. Furthermore, the results of the modelling analysis in Sect. 3.5.4 suggest that the protocol provides modelling resistance and tag unclonability.

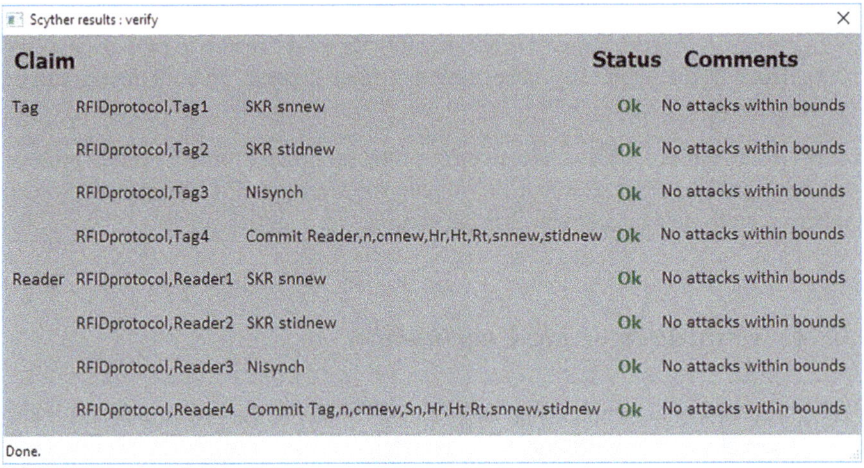

Scyther results : verify

| Claim | | | Status | Comments |
|---|---|---|---|---|
| Tag | RFIDprotocol,Tag1 | SKR snnew | Ok | No attacks within bounds |
| | RFIDprotocol,Tag2 | SKR stidnew | Ok | No attacks within bounds |
| | RFIDprotocol,Tag3 | Nisynch | Ok | No attacks within bounds |
| | RFIDprotocol,Tag4 | Commit Reader,n,cnnew,Hr,Ht,Rt,snnew,stidnew | Ok | No attacks within bounds |
| Reader | RFIDprotocol,Reader1 | SKR snnew | Ok | No attacks within bounds |
| | RFIDprotocol,Reader2 | SKR stidnew | Ok | No attacks within bounds |
| | RFIDprotocol,Reader3 | Nisynch | Ok | No attacks within bounds |
| | RFIDprotocol,Reader4 | Commit Tag,n,cnnew,Sn,Hr,Ht,Rt,snnew,stidnew | Ok | No attacks within bounds |

Done.

**Fig. 3.4** Scyther verification results

The security analysis was carried out with the Scyther tool to verify that the authentication protocol is resistant against the following attacks:

- An attacker in the *MitM(Man in the middle) attack* is positioned between legitimate parties controlling the authentication protocol. The attacker could compromise the authentication protocol security by exploiting the possibility of changing the content of messages between honest parties seized in the middle. As proven by Scyther that the *Nisync*, *commit* and *SKR* (*secret*) claims are satisfied, the attacker cannot break the mutual authentication with this kind of attack in the proposed protocol. Therefore, tag and reader mutually have an agreement over the exchanged messages with regard to their order and content.
- *An eavesdropping attack* is not possible due to the fact that all unprotected communications are either random or public data and all sensitive data is encrypted. Furthermore, maintaining that the response $s_{n_{\mathrm{new}}}$ and private key $q$ are held secret between the tag and the reader ensures that the content of exchanged messages cannot be decrypted.
- *A replay attack* is not possible. The signature of the tag's temporary ID $s_{tid}$, the message signature $H$t and the encrypted message $C$ are all randomised for each tag and each transaction, and any part of any communication cannot be partially reused.
- *A tracking attack* is also not possible since all data transmitted is either encrypted ($C$) or random ($R_{\mathrm{t}}$, $H_{\mathrm{t}}$, $H_{\mathrm{r}}$, $c_{n_{\mathrm{new}}}$). In addition, private and public keys are dynamic (i.e. new and different keys are generated after each session). Thus, the tag cannot be traced.
- *A desynchronisation attack* can be efficiently mitigated by the proposed protocol. Checking $\Delta k$ during the protocol process not only allows the reader to detect a desynchronisation attack but also allows the reader to retrieve as much secret information as needed in order to constantly maintain the required error margin.

- *Reader impersonation* is not possible because a fake reader cannot identify the tag from $s_{tid}$. Furthermore, it is not possible for a fake reader to provide a correct signature of the public key $s_n$ because that value depends on both the tag and the transaction.

All of the above validates the security of the proposed protocol against protocol attacks. The following section will validate the security of the proposed protocol against the physical attack, which is tag cloning.

### *3.5.4 Verification of Tag Unclonability*

The proposed tag design is enhanced with PUF technology, which makes it highly resistant to physical cloning attacks. In fact, the only way to clone the tag is to clone the PUF. As mentioned in [25], there are a number of security attacks that aims to clone a PUF, where these can be classified into two types, the first is physical cloning attacks, which have been demonstrated to be extremely difficult. The second type is based on the use of machine learning algorithms, typically referred to as modelling attacks, wherein attackers are assumed to be able to collect a large number of challenge/response pairs (CRP) set by listening to the communication channel [26]. Then, using this set, they attempt to obtain the full CRP sets of the PUF by a numerical model trained with machine learning methods (e.g. a support vector machine and a neural network). A reasonable approach to prevent such an attack is to conceal the relationship between the pairs of inputs and outputs using cryptosystems [27]. For example, Che et al. [28] employed a cryptic hash function applied to the input of the PUF. Cryptographic primitives, i.e. Hash function, encryption algorithms and XOR function can be used to hide the relationship between input and output. The challenge permutation and substitution techniques proposed in [29] are lightweight alternatives to make modelling attacks more difficult.

As an alternative to all of the aforementioned methods, the PUF responses shown in Fig. 3.5 have been hidden as follows: the response $s_{n_{\text{new}}}$ and $s_{tid_{\text{new}}}$ produced by the PUF was encrypted using the Rabin scheme. The encrypted output $C$ is generated by $C \leftarrow \text{RabinEnc}(M, n)$. This prevents any challenge-response sets from being collected from the communication channel by an attacker. Thus he/she cannot create a model for the PUF. In the following two sections, the security of the proposed authentication protocol, which uses the response-hiding method mentioned above, is assessed by employing ML attacks (support vector machine and neural network). The resistance of arbiter-PUF against ML attacks is tested for comparison.

#### 3.5.4.1 Test Vector Generation and Machine Learning

In the initial stage of the model building analysis, the challenge-response sets already utilised within the authentication process have been collected. After the

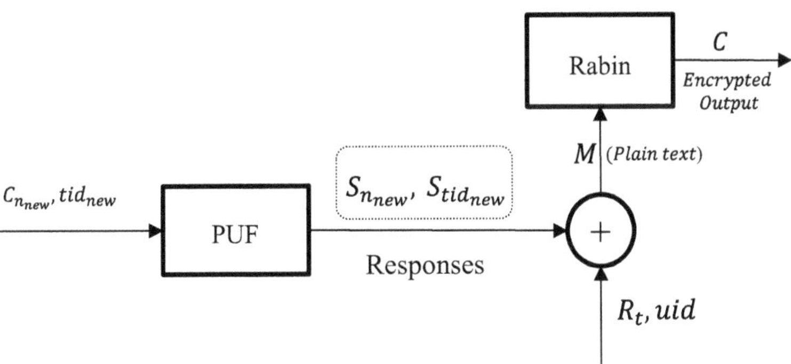


**Fig. 3.5** The response-hiding scheme using the Rabin algorithm

collection has been completed, model building attacks based on machine learning techniques are then constructed using these sets. In the traditional PUF authentication protocol described in [30], the verification is performed by explicitly transferring the input and output of the PUF between the prover and the verifier. Thus, the model building attack is a threat to such a protocol because the adversary may build a model obtaining enough number of input and output pairs as reported in [30]. Regarding the model building analysis of the proposed protocol; firstly, 32,000 test vectors are obtained from Arbiter-PUF. Then, 32,000 pairs of the encrypted output and the input are produced as a test vector from the scheme shown in Fig. 3.5. Subsequently, modelling those test vectors with NN and SVM is attempted. In the next section, the test results of SVM and NN over test vectors collected from the Arbiter PUF and the response-hiding scheme are evaluated.

#### 3.5.4.2 Model-Building Results

The SVM and ANN machine learning techniques were performed to examine the strength of the arbiter and the response-hiding scheme described in Fig. 3.5. Arbiter PUF was presented in [31]. In this chapter, the PUF takes the challenge as a 32-bit input to produce an output as a 1-bit response. During the modelling analysis, $b_i$ refers to the bits of the challenge ($C$), (i.e. $C = b_1, b_2, b_3, \ldots, b_k$) and $R$ interprets the output of the PUF being a response. In addition, the term $o_i$ refers to bits of the encrypted output ($C$), that is, $C = o_1, o_2, o_3, o_4, o_5, \ldots, o_m$. The value of $k$ is generally considered to be the bit length of the PUF. As shown in Fig. 3.6, the value of the encrypted 1024 bit output is generated by the Rabin encryption. Accordingly, the value of $m$ for the response-hiding scheme is 1024.

For comparative evaluation, the strength of the 32-bit arbiter facing ML attacks is tested. Figure 3.6 shows the predicting results of the ML attacks on the arbiter PUF. As can be seen from Fig. 3.6, ML techniques on the arbiter PUF have very high prediction accuracies of 99.5% (NN) and 98.4% (SVM). Then the resilience

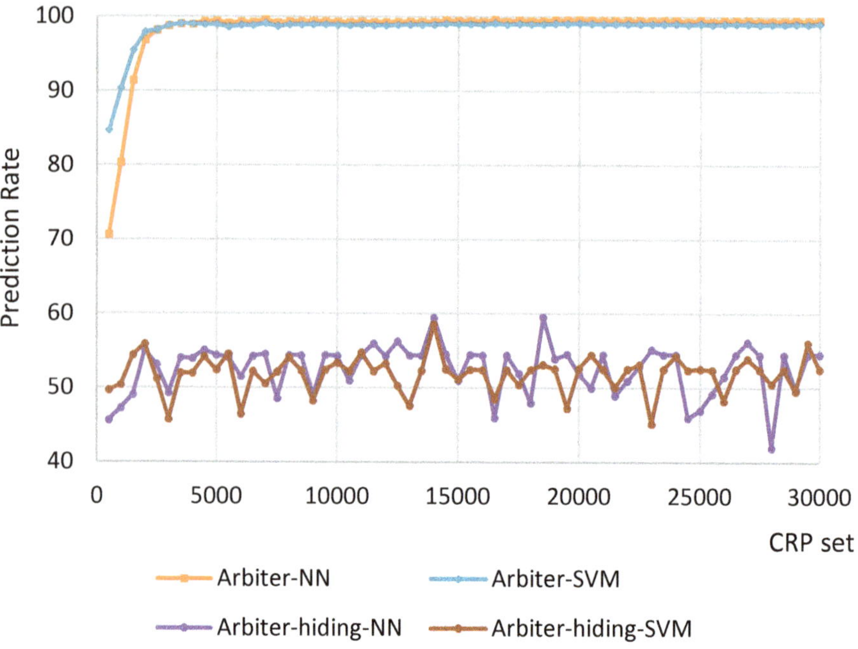


**Fig. 3.6** ML-attacks (SVM, ANN) on Arbiter PUF and Arbiter response-hiding scheme

of the response-hiding scheme for the arbiter was tested as well applying the same attacks. Figure 3.6 shows the prediction rate results according to both the SVM and NN techniques. The findings are based on 32,000 number of $(C_i)$ challenges and the first 5 bits of the encrypted output $o_i$, i.e. $(o_1, o_2, o_3, o_4, o_5)$. The results from such analysis revealed that the average prediction rate was 52.6% in NN and 51.9% in SVM for all $(C_i)$ and $(o_i)$ bit set. These results highlight that the prediction rate could not be increased by getting further $(C_i)$ and encrypted output $(o_i)$ sets. It is necessary to note that the same machine learning attacks on the individual output bits of $o_i$ can be employed for analysis of the entire output of the response masking scheme, as in the Arbiter analysis. The finding from the analysis shows that the probability of one-bit predicting is 1/2 for the response masking scheme shown in Fig. 3.5. The odds of predicting the entire bits of the encrypted output must then be $1/2^m(1/2^{1024})$ in a trial. Consequently, if an attacker attempts to decode a 1024-bit key-coded encrypted output, he or she must perform approximately $2^{1024}$ computations to obtain the plain text. For RFID tag devices using the public-key-based algorithm, the 1024-bit security levels support sufficient protection currently, as mentioned in [20]. Furthermore, the ML test results show that the difficulty of predicting the entire output of the response-hiding scheme is surely increased exponentially by the wrong estimation of one-bit $o_i$. The results also indicate that the latency of the response guarantees bit security. Consequently, the introduced response masking technique shows a clear resilience against machine learning attacks.

So far, the analysis presented above validates the security of the proposed protocol against protocol attacks and model building attacks. In the following section comparison with other protocols will be given considering the security attacks and the security requirements of the proposed protocol.

### 3.5.5 Comparison with Related Work

The security requirements of the proposed protocol and its capabilities to resist attacks are compared with other authentication protocols in Table 3.2. In this section, the proposed protocol is compared with the other authentication protocols such as WIPR design [5], a PUF-based protocol recently proposed in [8] and an ECC and PUF-based protocol recently proposed in [9].

WIPR [5] is an authentication protocol based on the low-cost Rabin scheme and tag hardware design. This authentication protocol is not complicated and does not support mutual authentication. Moreover, all-the tags share the same public key, and this protocol itself cannot protect the system from tag cloning attacks. The proposed protocol overcomes these drawbacks by using PUF intensively along with introducing mutual authentication and public-key transmission into the protocol. However, the WIPR protocol which has no ability to support mutual authentication might consume less energy and require less area than the proposed protocol.

In term of security, the two protocols reported in [8, 9] are comparable. The main difference is the way the tags identify themselves to the reader, where the tags in [9] transmit their UID in the unprotected channel at the beginning of the transaction, while the protocol in [8] uses TID instead. The former approach [9] suffers from tracking attacks and violates confidentiality and anonymity requirements, while the latter [8] is unable to mitigate tracking attacks against the TID, and also needs to synchronise shared knowledge which introduces potential availability issues.

The proposed protocol uses TID just like [8], but the solution for the availability problem is complete and more efficient. Note that although Gope et al. [8] claimed that their protocol satisfied the availability requirements, they only partially resolve the problem. First, they use a set of emergency CRPs stored in the readers and a set-of-unique unlinkable pseudo-identities stored in the tag in order to recover from a desynchronisation-attack. The proposed protocol does not require any CRP storage in the tag as it only requires a CRP storage in the reader, where resources are considered to be unlimited. Secondly, their protocol does not provide any approach to refill the emergency sets after being attacked by an adversary. The proposed protocol, however, allows the reader to maintain a number of transactions' secret information and allows the reader to obtain more secrets from the tag after any successful authentication.

The above comparison leads to a conclusion that previous works have been limited to providing partial protection for RFID systems. In contrast, this chapter set out with the aim of proposing a new security protocol, which simultaneously provides lightweight mutual authentication, availability and tag unclonability in the

RFID system. The following section describes the tag design and implementation in the context of the security primitives of the proposed protocol.

## 3.6 Design and Implementation

This section outlines the cryptographic primitives used in the design and implementation of the proposed protocol by considering the tag side as the tag in the RFID system is deemed to be the resource-constrained device in the IoT domain. The evaluation of the used cryptographic primitives in terms of implementation cost will be discussed, and the statistical characteristics of the proposed protocol arising from the used cryptographic primitives will be addressed in Sect. 3.7.

### 3.6.1 Tag Design

The design of the tag is presented in this section. Because the proposed protocol is developed based on the WIPR design (which was proposed in [5]), the Rabin encryption is implemented in the same way. The data path of the WIPR encryption is presented in Fig. 3.7.

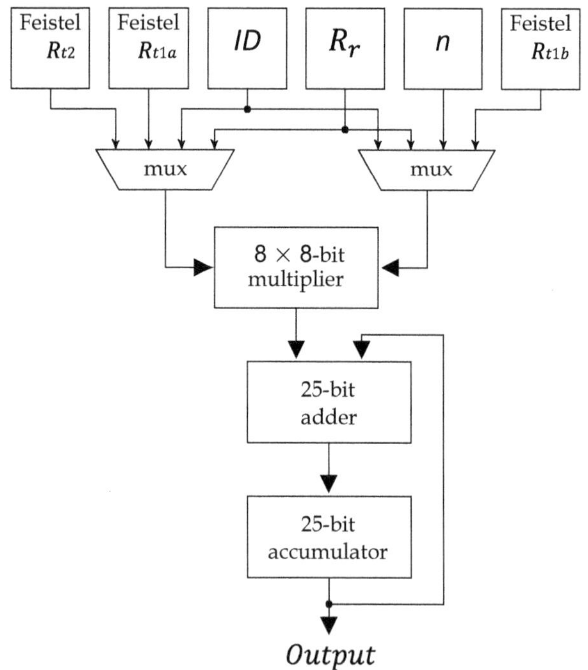


**Fig. 3.7** Data path for the WIPR Rabin encryption [5]

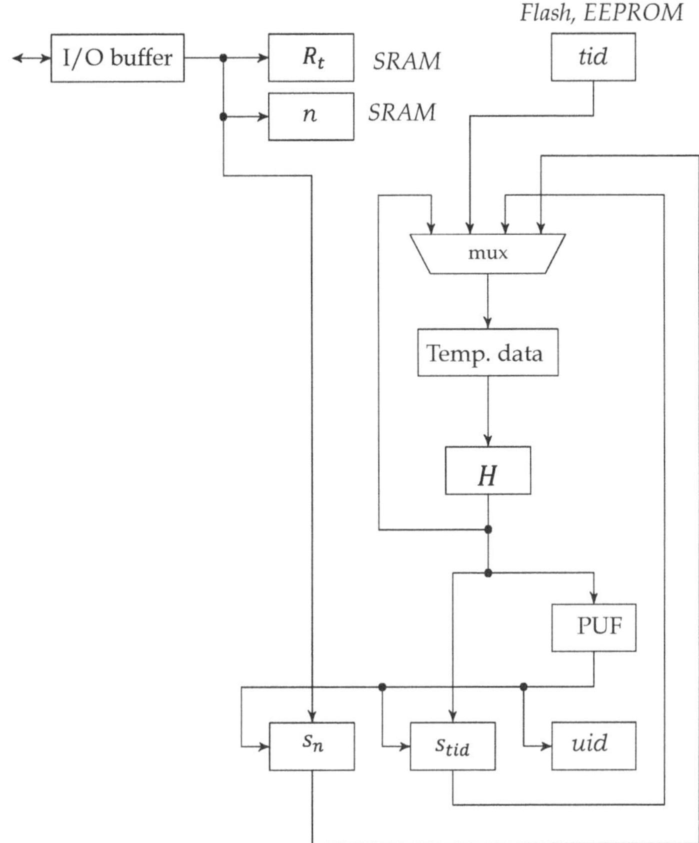


**Fig. 3.8** The design of the security system of the RFID tag for the proposed protocol

In order for the tag to perform the proposed protocol, some additional security primitives and registers (listed below) are added. The data-path of the extended-circuit is presented in Fig. 3.8.

- Memories and registers
  - Public-key memory $n$: 1024-bit single-port SRAM
  - Random value $R_r$: 80-bit dual-port SRAM
  - Temporary ID *tid*: 80-bit non-volatile memory (NAND flash, EEPROM)
  - $s_n$, $s_{tid}$, $uid$ and temporary register for $H$: 80-bit registers
- Security primitives
  - Hash: Hash-One
  - PUF: 80-bit Arbiter PUF

All of the above security primitives, memories and registers are used in the detailed hardware design of the security system of the RFID tag. How they are employed in the protocol process is outlined in the following subsections.

### 3.6.2 Hash Function (H)

The hash function is used to generate the temporary ID, challenge sequence for the PUF and message signatures $H_t$, $H_r$, $H$. This function is not necessarily to be cryptographically strong because it is not directly used for security purpose. Therefore, a simple and area-efficient design is preferable. In the proposed protocol design, the lightweight cryptographic hash function proposed in [32] is used.

### 3.6.3 PUF

In the proposed protocol, an A-PUF described in [33] is used as follows. The A-PUF takes a 40-bit challenge and generates a 1-bit response. Therefore, in order to generate an 80-bit output, the PUF is fed with a sequence of challenges generated by the hash function $H$.

### 3.6.4 Memories and Registers

The public key $n$ in the Rabin cryptosystem has a large size and strongly affects the area and power consumption of the tag. But, the public key is always randomly accessed by address and consequently can be put into an SRAM which is more area-efficient than a register. The random value $R_t$ can also be stored in an SRAM. The current temporary ID *tid* must be stored in non-volatile memory, such as a NAND flash or an EEPROM.

All $s_n$, $s_{tid}$, *uid* values and the hash function temporary data are 80-bit and are required to be stored in registers. They are the largest part of the design. Using SRAM is also possible for $s_n$, $s_{tid}$ and *uid*; however, this would require different and more complicated design and control.

The following section will evaluate the area-related cost of all of the above primitives' hardware implementations and statistical characteristics of the proposed protocol resulting from the PUF cryptographic primitive.

## 3.7 Evaluation and Cost Analysis

Before interpreting cost analysis results, the reader is reminded of the main aim of this chapter which is to find a lightweight and secure solution to mutual authentication using PUF and Rabin encryption for the RFID system. In this section, the statistical characteristics of the proposed authentication protocol and the tag implementation cost are discussed.

### *3.7.1 Statistical Characteristics*

There are two crucial statistical metrics for an authentication protocol: False Acceptance Rate (FAR) and False Rejection Rate (FRR), which are outlined as follows:

- FAR is the probability of the authentication process completing successfully although either the tag or the reader is not authentic.
- FRR is the probability of the authentication process terminating unsuccessfully although both the tag and the reader are authentic.

In practice, the goal is to minimise both FAR and FRR. Since FAR and FRR are both monotonic functions of threshold value with the opposite monotonicity, they reach the lowest value when they are equal. That value is called the Equal Error Rate (ERR).

#### 3.7.1.1 PUF Performance Metrics

The reliability of the authentication process depends closely on the statistical characteristics of the PUF. Therefore, fundamental performance metrics of PUFs, namely uniqueness and reliability, are mentioned as follows.

- *Uniqueness* is the ability of PUFs in different devices to generate a unique response for the same challenge. It is represented as the average inter-chip Hamming distance of the $n$-bit responses $R_i(n)$ and $R_j(n)$ generated by two different chips $i$ and $j$, respectively:

$$\mathrm{HD_{inter}} = \frac{2}{k\left(k-1\right)} \sum_{i=1}^{k-1} \sum_{j=i+1}^{k} \frac{\mathrm{HD}\left(R_i(n), R_j(n)\right)}{n} \times 100\% \tag{3.17}$$

Ideally, the inter-chip Hamming distance of a PUF is 50%. For a detailed explanation of formula (3.17), readers are referred to Chap. 2 in [25].

- *Reliability* is the ability of PUFs in the same device to generate the same response for the same challenge. It is represented as the average intra-chip Hamming distance of the $n$-bit responses $R_i(n)$ and $R_i'(n)$ generated by the chip $i$ in different conditions:

$$\mathrm{HD}_{\mathrm{intra}} \frac{1}{k} \sum_{i=1}^{k} \frac{\mathrm{HD}\left(R_i(n), R_i'(n)\right)}{n} \times 100\% \tag{3.18}$$

Ideally, the intra-chip Hamming distance of a PUF is 0%. For a detailed explanation of formula (3.18), readers are referred to Chap. 2 in [25].

#### 3.7.1.2 Performance of Single PUF Authentication

As described in [34], the performance analysis of the basic PUF-based protocol is carried out as follows. Let $t$ be the acceptable number of error bits in the PUF response and assume that $\mathrm{HD}_{\mathrm{inter}}$ and $\mathrm{HD}_{\mathrm{intra}}$ of PUFs follow binomial distribution with the binomial probability estimators of $p_{\mathrm{inter}}$ and $p_{\mathrm{intra}}$, respectively. Furthermore, let $\overline{A}(t)$ and $\overline{R}(t)$, respectively, be the FAR and FRR of an $n$-bit response PUF with the error rate threshold $t$. FAR, denoted with $\overline{A}(t)$, and FRR, denoted with $\overline{R}(t)$, can be calculated as follows [34]:

$$\overline{A}(t) = \sum_{i=0}^{t} \binom{i}{n} p_{\mathrm{inter}}^{i} (1 - p_{\mathrm{inter}})^{n-i} \tag{3.19}$$

$$\overline{R}(t) = 1 - \sum_{i=0}^{t} \binom{i}{n} p_{\mathrm{intra}}^{i} (1 - p_{\mathrm{intra}})^{n-i} \tag{3.20}$$

As shown in Fig. 3.9, it is obvious that $\overline{A}(t)$ is a monotonically increasing function and $\overline{R}(t)$ is a monotonically decreasing function. The stricter the acceptance threshold $t$ is, the harder it is to accept a wrong tag and the more likely it is to reject the right tag.

#### 3.7.1.3 Performance of the Proposed Authentication Protocol

In the proposed protocol described in Sect. 3.4.5, successful authentication requires a matching temporary ID donated by $s_{tid}$, a matching public-key signature donated by $s_n$ and a matching unique ID donated by $uid$, which are all PUF responses. However, the protocol accepts the collision of $s_{tid}$ and overcomes such issue by

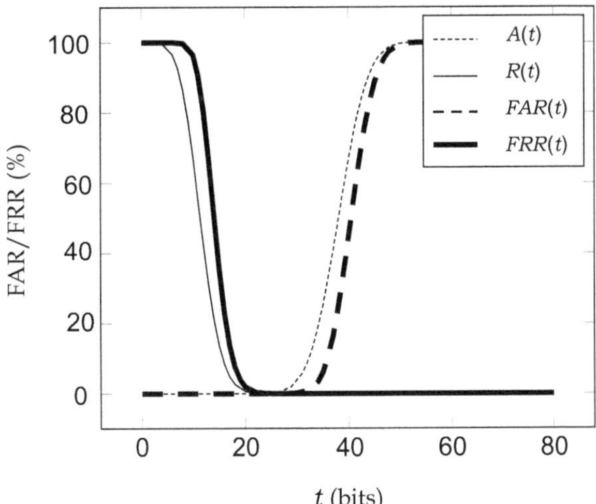


**Fig. 3.9** FAR and FRR of the Arbiter PUF and the proposed authentication protocol, with $p_{\text{intra}} = 15\%$ and $p_{\text{inter}} = 48.2\%$ over different threshold values

trying all the tags, which have a matched $s_{tid}$. Therefore $s_{tid}$ does not contribute to FAR.

FAR of the proposed authentication protocol is calculated as follows:

$$\text{FAR}(t) = \overline{A}^2 \tag{3.21}$$

During the execution of the proposed protocol, an authentication process could be incorrectly terminated at either the $s_{tid}$, $s_n$ or *uid* checking steps. These values are all generated by the PUF, and therefore the probabilities of each case are $\overline{R}$, $\left(1 - \overline{R}\right)\overline{R}$ and $\left(1 - \overline{R}\right)^2\overline{R}$, respectively.

FRR of the proposed authentication protocol is calculated as follows:

$$\text{FRR}(t) = \overline{R} + \left(1 - \overline{R}\right)\left[\overline{R} + \left(1 - \overline{R}\right)\overline{R}\right] \tag{3.22}$$

The FAR and FRR of the proposed authentication protocol have the same monotonicity as the FAR and FRR of the PUF. The goal is to choose the threshold value $t_{\text{EER}}$ such that both FAR and FRR are minimum as shown in Fig. 3.9.

The threshold value $t_{\text{EER}}$ is calculated as follows:

$$t_{\text{EER}} = \arg\min_t \left[\max\left(\text{FAR}(t), \text{FRR}(t)\right)\right] \tag{3.23}$$

in this case, the equal error rate is

$$\text{EER} = \max\left(\text{FAR}(t), \text{FRR}(t)\right) \tag{3.24}$$

Based on the above calculations, Fig. 3.9 shows the FAR and FRR of the Arbiter PUF and the proposed authentication protocol using that PUF. When the PUF has

$p_{\text{intra}} = 15\%$ and $p_{\text{inter}} = 48.2\%$, the preceding performance analysis leads to the determining that the threshold value $t$ could be chosen in the range of 26–29. This means that if the threshold $t_{\text{EER}}$ is selected within the specified range, FAR and FRR of the proposed protocol are eliminated effectively as shown in Fig. 3.9.

## *3.7.2 Cost Analysis*

In this section, the area-related cost of the tag involved in the proposed protocol is estimated as a resource-constrained party, and the estimation results are then evaluated by comparing them with other existing protocols.

### 3.7.2.1 Area of the Tag

The tag design is synthesised using CMOS 0.35 μm technology. The area of the 1024-bit tag design and those of some important components are estimated in gate equivalents (GEs) and presented in Table 3.3.

As can be seen in Table 3.3, the complete tag design is 10,139 GEs which is acceptable for a lightweight passive RFID tag. Most of the area in the tag is occupied by registers and memories (over 60%), while a considerable amount of area is occupied by multiplexers and control circuits (over 20%). Security primitives, namely PUF and hash function, occupy relatively small area, as is indicated by the data in Table 3.3.

The following valuable optimisations could improve the security and the area consumption of the tag:

**Table 3.3** Total area of the 1024-bit tag and areas of the constituent important components

| Components | Area (GEs) | Area (%) |
|---|---|---|
| I/O buffer 8-bit register | 48 | 0.5 |
| $s_n$, $s_{tid}$, $uid$, PRNG state 80-bit registers | 2188 | 21.6 |
| $R_{\text{t1a}}$, $R_{\text{t1b}}$, $R_{\text{t2}}$ 80-bit registers | 1440 | 14.2 |
| 1024-bit key SRAM | 1536 | 15.1 |
| 80-bit $R_{\text{r}}$ SRAM | 120 | 1.2 |
| A-PUF 80-bit | 326 | 3.2 |
| Hash function | 1006 | 9.9 |
| Feistel logic | 649 | 6.4 |
| 25-bit adder | 113 | 1.1 |
| 8 × 8-bit multiplier | 394 | 3.9 |
| 25-bit accumulator | 150 | 1.5 |
| Others | 2169 | 21.4 |
| Total area | 10,139 | 100 |

- If the area of the registers could be optimised by using fewer registers and using SRAM as a replacement, the total area of the design could be reduced significantly since most of the design area is occupied by memories and registers.
- PUF is the vital security primitive in the design, and as such, more area should be dedicated on a more sophisticated PUF design.

The above estimation results validate the lightweight nature of tag design for the proposed protocol. Based on those estimations, a comparison with other authentication protocols will be presented in the next section.

#### 3.7.2.2 Comparison with Other Authentication Protocols

In this section, the same authentication protocols as those discussed in Sect. 3.5.5 will be compared in term of design cost to the proposed protocol. Each study provides the design cost information in a different way, which makes it much harder to have an accurate total area unless all tag designs are re-implemented entirely. Furthermore, each design may use different implementation for the same security function, such as TRNG or a hash function. A summary of the required security primitives and amount of memories of various protocols considered here is given in Table 3.4.

In order to have a brief view of the design cost, the area of the tags in each protocol will be estimated based on the following assumptions:

- All protocols are implemented using the same security primitives provided in Table 3.5. Note that the area of the Rabin encryption excludes all memories (which are counted separately, see Table 3.4) and the one-way function (which is considered equivalent to PRNG).

**Table 3.4** Comparison between the resources for the tag required by proposed protocol and other authentication protocols

| Protocols | [5] | [8] | [9] | Proposed |
|---|---|---|---|---|
| *Components* | | | | |
| TRNG | ● | ● | ● | ● |
| Hash | ○ | ● | ● | ● |
| PRNG | ●[a] | ● | ○ | ○ |
| PUF | ○ | ● | ● | ● |
| ECC | ○ | ○ | ● | ○ |
| Rabin encryption | ● | ○ | ○ | ● |
| *Memories* | | | | |
| Non-volatile (bits) | 1024 | $128 + 64 \times k^{b}$ | –[c] | 80 |
| Registers (bits) | 368 | 576 | –[c] | 568 |
| SRAMs (bits) | 0 | 0 | –[c] | 1104 |

●: required, ○: not required
[a]Assume that PRNG is equivalent to the one-way function
[b]$k$ is the number of emergency unlinkable pseudo-identities
[c]Not provided

**Table 3.5** Area of the security primitive hardware implementations

| Primitives | Type | Name | Security strength | Area (GEs) |
|---|---|---|---|---|
| [35] | TRNG | TRNG | – | 72 |
| [32] | Hash function | Hash-One | 80 | 1006[a] |
| [36] | PRNG | xorshift+ | 80 | 383 |
| [33] | PUF | A-PUF 80-bit | – | 326 |
| [37] | Encryption | ECC 163-bit | 80[b] | 12,145 |
| [5] | Encryption | Rabin 1024-bit | 80[b] | 1733[c] |

[a] Serial variant (smaller area, more cycles)
[b] Comparable strength recommended by NIST [20]
[c] Excluding payload memory and one-way function (counted later as part of the system)

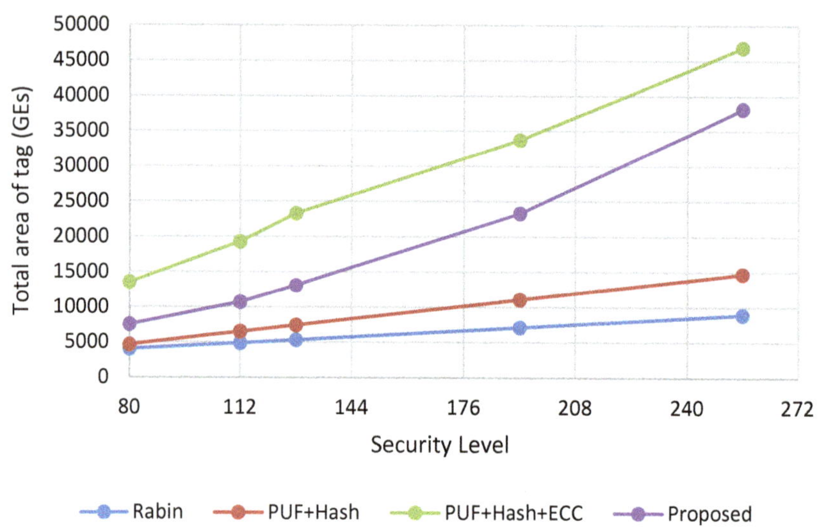


| Security Level | Rabin[5] | PUF+Hash[8] | PUF+Hash+ECC[9] | Proposed |
|---|---|---|---|---|
| **80** | 4028 | 4667 | 13549 | 7513 |
| **112** | 4917 | 6505 | 19297 | 10717 |
| **128** | 5361 | 7424 | 23289 | 13088 |
| **192** | 7140 | 11100 | 33743 | 23337 |
| **256** | 8918 | 14776 | 46879 | 38195 |

**Fig. 3.10** The estimated total area (GE) of the tags in each authentication protocol for different security level

- The memories bits and the areas of all security primitives except TRNG and Rabin encryption are proportional to their security strength/state bits/challenge bits/key length.

The tag area is estimated for different security strengths from 80 to 256 as shown in Fig. 3.10. The key lengths of ECC and Rabin encryption equivalent to each security strength are chosen based on the NIST recommendation [20]. The area of non-volatile memories is considered as negligible, whereas the area of registers and SRAMs is approximately 6 and 1.5 GE/bit, respectively, as can be inferred from Fig. 3.10.

The above analysis of area estimations leads to the conclusion that:

- The proposed protocol requires more tag area than the protocols based on only Rabin encryption [5] and on PUF and hash function [8] due to the large memory required for the public key and the Rabin encryption block. This is acceptable because, unlike the proposed protocol, neither of the protocols proposed in [5, 8] offers a complete security solution in the context of the three qualities of lightweight mutual authentication, availability and tag unclonability, as discussed in Sect. 3.1.
- The proposed protocol requires a smaller area for the tag (20–50%) than the protocol based on PUF and ECC which contains the highly complex ECC block. The larger key length of Rabin compared to ECC forces the tag area of the proposed protocol to grow faster than in the ECC-based design. However, the proposed protocol is still 20% more area-efficient at the highest security strength (Rabin key length of 15,360 bits) as can be seen in Fig. 3.10.

## 3.8 Conclusion

The use of RFID technology is a very promising solution to mitigate the risk of counterfeiting. To achieve this, RFID systems need to be resilient against known security threats including tag cloning attacks. Building a secure RFID system is a challenging task due to their limited computing resources. This chapter has proposed a lightweight authentication protocol along with secure hardware design for the tag. The proposed solution combines Rabin public-key cryptosystem and PUF technology to perform mutual authentication and public-key transmission. The security of the proposed scheme has been systematically analysed and proved against known attacks, including man in the middle, eavesdropping, replay, tracking, reader impersonation, desynchronisation; in addition, the resilience of the proposed tag design against modelling attacks has been experimentally demonstrated. The evaluation results show that the proposed protocol is up to 50% more area-efficient compared with Elliptic-curve-based schemes.

## A.1 Appendix: Sample Code for the Proposed Protocol

```
// Project : proposed protocol
// Program name : main.c
// Author : yy6e14
// Date created : 20/7/2018
```

```
// Purpose : In order to test authentication in RFID system
// Revision History :
// Date Author Ref Revision (Date in 01/8/2018 )
// 01/8/2017 yy6e14 1
#include <rfidauth/reader.h>
#include <rfidauth/tag.h>
#include <rfidauth/core.h>
#include <stdio.h>
#include <stdlib.h>
#include <time.h>
#include <pthread.h>
#include <semaphore.h>
#include <auth4.h>
#include <tommath.h>

static void auth_protocol_test(void);

static void auth_bigint_test(void);
static void auth_prng_test(void);

int main(void)
{
    srand(time(NULL));
    if (1) auth_protocol_test();
    if (0) auth_prng_test();
    return 0;
}

static auth_RFID_tag_t RFID_tag;
static auth_reader_t RFID_reader;
static pthread_t RFID_tag_thread;
static sem_t comm_RFID_tag_sem_count, comm_RFID_tag_sem_capacity;
static uint8_t comm_RFID_tag_buffer;
static sem_t comm_RFID_reader_sem_count, comm_RFID_reader_sem_capacity;
static uint8_t comm_RFID_reader_buffer;

static void auth_protocol_test_send(void *sender, const uint8_t *buffer, int len)
{
    int i;
    if (sender == &RFID_tag)
    {
        printf("RFID_tag_sent:");
        for (i = 0; i < len; ++i)
          printf("0x%02x", buffer[i]);
      printf("\n");

        for (i = 0; i < len; ++i)
```

```
            {
                sem_wait(&comm_RFID_reader_sem_capacity);
                comm_RFID_reader_buffer = buffer[i];
                sem_post(&comm_RFID_reader_sem_count);
            }
        }
        else if (sender == &RFID_reader)
        {
            printf("RFID_reader_sent:");
            for (i = 0; i < len; ++i)
                printf(" 0x%02x", buffer[i]);
            printf("\n");

            for (i = 0; i < len; ++i)
            {
                sem_wait(&comm_RFID_tag_sem_capacity);
                comm_RFID_tag_buffer = buffer[i];
                sem_post(&comm_RFID_tag_sem_count);
            }
      }
}
static void auth_protocol_test_RFID_reader_recv(void *receiver, uint8_t *buffer,
   int len)
{
        int i;

        for (i = 0; i < len; ++i)
        {
              sem_wait(&comm_RFID_reader_sem_count);
              buffer[i] = comm_RFID_reader_buffer;
              sem_post(&comm_RFID_reader_sem_capacity);
        }
}
static void auth_protocol_test_RFID_tag_recv(void *receiver, uint8_t *buffer, int
   len)
{
        int i;

        for (i = 0; i < len; ++i)
        {
              sem_wait(&comm_RFID_tag_sem_count);
              buffer[i] = comm_RFID_tag_buffer;
              sem_post(&comm_RFID_tag_sem_capacity);
        }
}

static void auth_protocol_test(void)
```

```
{
        uint8_t packet_type, k;
        AUTH_BIGINT_DECLARE(data, AUTH_SECRET_BITS);

        AUTH_BIGINT_INIT(data, AUTH_SECRET_BITS);

        /* Initialize__RFID_tag_and__RFID_reader */
        auth_RFID_tag_init(&RFID_tag,
        auth_protocol_test_send, auth_protocol_test_RFID_tag_recv);
        auth_RFID_reader_init(&RFID_reader,
        auth_protocol_test_send, auth_protocol_test_RFID_reader_recv);

        /* Initialize_the_communication_channel */
        sem_init(&comm_RFID_tag_sem_count, 0, 0);
        sem_init(&comm_RFID_tag_sem_capacity, 0, 1);
        sem_init(&comm_RFID_reader_sem_count, 0, 0);
        sem_init(&comm_RFID_reader_sem_capacity, 0, 1);

        /* Start__RFID_tag_receiver_event_loop */
        pthread_create(&RFID_tag_thread, NULL,
        (void*)auth_RFID_tag_event_loop, &RFID_tag);

        /* Set_up_RFID_tag */
        auth_RFID_reader_setup_RFID_tag(&RFID_reader);

        /* Authenticate_RFID_tag */
        auth_RFID_reader_authenticate_RFID_tag(&RFID_reader);
        auth_RFID_reader_authenticate_RFID_tag(&RFID_reader);
        auth_RFID_reader_authenticate_RFID_tag(&RFID_reader);
        auth_RFID_reader_authenticate_RFID_tag(&RFID_reader);
        auth_RFID_reader_authenticate_RFID_tag(&RFID_reader);
}
static void auth_prng_test(void)
{
        int i;
        auth_prng_state_t state;
        AUTH_BIGINT_DECLARE(output, AUTH_SECRET_BITS / 2);

        auth_prng_init(&state, NULL, NULL);
        AUTH_BIGINT_INIT(output, AUTH_SECRET_BITS / 2);

        printf("%s", "–Random_test_PRNG\n");

        auth_bigint_urandom(&state.lo, AUTH_SECRET_BITS / 2);
        auth_bigint_urandom(&state.hi, AUTH_SECRET_BITS / 2);

        for (i = 0; i < 100; ++i)
        {
              auth_prng_next(&state);
```

```
        auth_prng_output_half(&output, &state);

        printf("%s", "—PRNG_state = ("); auth_bigint_print_hex(&state.lo);
        printf("%s", ", "); auth_bigint_print_hex(&state.hi);
        printf("%s", ") => _output = "); auth_bigint_print_hex(&output);
        printf("\n");
    }
}
```

## References


1. O.V. Buowari, Fake and counterfeit drug: a review. Afrimedic J. **3**(2), 1–4 (2012)
2. P. Kitsos, Y. Zhang, *RFID Security*, vol 1(4) (Springer, Boston, MA, 2008)
3. K. Bu, M. Weng, Y. Zheng, B. Xiao, X. Liu, You can clone but you cannot hide: a survey of clone prevention and detection for RFID. IEEE Commun. Surv. Tutorials **19**(3), 1682–1700 (2017)
4. A. Riahi, E. Natalizio, Y. Challal, Z. Chtourou, A roadmap for security challenges in the Internet of Things. Digit. Commun. Networks **4**(2), 118–137 (2018)
5. A. Arbit, Y. Livne, Y. Oren, A. Wool, Implementing public-key cryptography on passive RFID tags is practical. Int. J. Inf. Secur. **14**(1), 85–99 (2014)
6. L. Fu, X. Shen, L. Zhu, J. Wang, A low-cost UHF RFID tag chip with AES cryptography engine. Secur. Commun. Networks **7**(2), 365–375 (2014)
7. D.M. Wang, Y.Y. Ding, J. Zhang, J.G. Hu, H.Z. Tan, Area-efficient and ultra-low-power architecture of RSA processor for RFID. Electron. Lett. **48**(19), 1185 (2012)
8. P. Gope, J. Lee, T.Q.S. Quek, Lightweight and practical anonymous authentication protocol for RFID systems using physically unclonable functions. IEEE Trans. Inf. Forensics Secur. **13**(11), 2831–2843 (2018)
9. U. Chatterjee et al., Building PUF based authentication and key exchange protocol for IoT without explicit CRPs in verifier database. IEEE Trans. Depend. Secur. Comput. **16**(3), 424–437 (2019)
10. D. He, S. Zeadally, An analysis of RFID authentication schemes for internet of things in healthcare environment using elliptic curve cryptography. IEEE Internet Things J. **2**(1), 72–83 (2015)
11. A. Braeken, PUF based authentication protocol for IoT. Symmetry (Basel) **10**(8), 352 (2018)
12. M.O. Rabin, *Digitalized Signatures and Public-Key Functions as Intractable as Factorization* (MIT Laboratory for Computer Science, Cambridge, MA, 1979)
13. C. Manifavas, G. Hatzivasilis, K. Fysarakis, K. Rantos, Lightweight cryptography for embedded systems—a comparative analysis, in *Lecture Notes in Computer Science*. Including Subseries Lecture Notes in Artificial Intelligence and Lecture Notes in Bioinformatics, vol. 8247, 2014, pp. 333–349
14. A. Shamir, Memory efficient variants of public-key schemes for smart card applications, in *Workshop on the Theory and Application of of Cryptographic Techniques*, 1995, pp. 445–449
15. B. Halak, M. Zwolinski, M.S. Mispan, Overview of PUF-based hardware security solutions for the internet of things, in *Midwest Symposium on Circuits and Systems*, Oct 2017, pp. 16–19
16. Y. Yilmaz, S.R. Gunn, B. Halak, Lightweight PUF-based authentication protocol for IoT devices, in *2018 IEEE 3rd International Verification and Security Workshop (IVSW)*, 2018, pp. 38–43
17. G.E. Suh, S. Devadas, Physical unclonable functions for device authentication and secret key generation, in *2007 44th ACM/IEEE Design Automation Conference*, vol. 129, 2007, pp. 9–14

18. A.P. Johnson, R.S. Chakraborty, D. Mukhopadhyay, A PUF-Enabled Secure Architecture for FPGA-Based IoT Applications. IEEE Trans. Multi-Scale Comput. Syst. **1**(2), 110–122 (2015)
19. J. Zhang, Y. Lin, Y. Lyu, G. Qu, A PUF-FSM binding scheme for FPGA IP protection and pay-per-device licensing. IEEE Trans. Inf. Forensics Secur. **10**(6), 1137–1150 (2015)
20. E. Barker, Recommendation for key management part 1: general, NIST Spec. Publ., Jan 2016, pp. 51–54
21. T. Idriss, H. Idriss, M. Bayoumi, A PUF-based paradigm for IoT security, in *2016 IEEE 3rd World Forum on Internet of Things (WF-IoT)*, 2016, pp. 700–705
22. M.N. Al Dalaien, S.A. Hoshang, A. Bensefia, A.R.A. Bathaqili, Internet of Things (IoT) security and privacy, in *Powering Internet Things with 5G Networks*, 2017, pp. 247–267
23. C.J.F. Cremers, The Scyther tool: verification, falsification, and analysis of security protocols, in *Computer Aided Verification*, LNCS, vol. 5123, (Springer, Berlin, 2008), pp. 414–418
24. D. Mendez Mena, I. Papapanagiotou, B. Yang, Internet of things: survey on security. Inf. Secur. J. **27**(3), 162–182 (2018)
25. B. Halak, *Physically Unclonable Functions From Basic Design Principles to Advanced Hardware Security Applications*, 1st edn. (Springer International Publishing, Cham, 2018)
26. D. Mukhopadhyay, PUFs as promising tools for security in internet of things. IEEE Des. Test **33**(3), 103–115 (2016)
27. M. Barbareschi, P. Bagnasco, A. Mazzeo, Authenticating IoT devices with physically unclonable functions models, in *2015 10th International Conference on P2P, Parallel, Grid, Cloud and Internet Computing (3PGCIC)*, 2015, pp. 563–567
28. W. Che, PUF-based authentication invited paper, in *IEEE/ACM International Conference on Computer-Aided Design*, 2015, pp. 337–344
29. M.S. Mispan, H. Su, M. Zwolinski, B. Halak, Cost-efficient design for modeling attacks resistant PUFs, in *Proceedings of the 2018 Design, Automation and Test in Europe Conference and Exhibition, DATE 2018*, vol. 2018, Jan 2018, pp. 467–472
30. J. Delvaux, R. Peeters, D. Gu, I. Verbauwhede, A survey on lightweight entity authentication with strong PUFs. ACM Comput. Surv. **48**(2), 1–42 (2015)
31. B. Halak, Y. Hu, M.S. Mispan, Area efficient configurable physical unclonable functions for FPGAS identification, in *Proceedings of the IEEE International Symposium Circuits and Systems*, vol. 2015, 2015, pp. 946–949
32. S. Manayankath, C. Srinivasan, M. Sethumadhavan, P. Megha Mukundan, Hash-One: a lightweight cryptographic hash function. IET Inf. Secur. **10**(5), 225–231 (2016)
33. J.W. Lee, B. Daihyun Lim G.E. Gassend M. van Dijk Suh, S. Devadas, A technique to build a secret key in integrated circuits for identification and authentication applications, in *2004 Symposium on VLSI Circuits. Digest of Technical Papers* (IEEE Cat. No. 04CH37525), 2004, pp. 176–179
34. Y. Gao, H. Ma, D. Abbott, S.F. Al-Sarawi, PUF sensor: exploiting PUF unreliability for secure wireless sensing. IEEE Trans. Circuits Syst. I Regul. Pap. **64**(9), 2532–2543 (2017)
35. P.Z. Wieczorek, K. Golofit, True random number generator based on flip-flop resolve time instability boosted by random chaotic source. IEEE Trans. Circuits Syst. I Regul. Pap. **65**(4), 1279–1292 (2018)
36. S. Vigna, Further scramblings of Marsaglia's xorshift generators. J. Comput. Appl. Math. **315**, 175–181 (2017)
37. X. Tan, M. Dong, C. Wu, K. Ota, J. Wang, D.W. Engels, An energy-efficient ECC processor of UHF RFID tag for banknote anti-counterfeiting. IEEE Access **5**, 3044–3054 (2017)

# Part III
# Anomaly Detection in Embedded Systems

# Chapter 4
# Anomalous Behaviour in Embedded Systems

**Lai Leng Woo**

## 4.1 Introduction

Embedded systems are becoming more common and widely used in various applications and devices such as automotive industry, factory automation, medical and health, power plants, telecommunication, smart homes, robotics and many others [2, 24]. Driven by advances in microelectronics and software, embedded systems are becoming more affordable for daily usage and have, thus, enriched our lives and connected people together. According to the new market research report on embedded systems market [23], this market is projected to grow from USD 86.5 billion in 2020 to USD 116.2 billion by 2025, mainly driven by the increase in the number of research and development activities related to embedded systems, rise in demand for advanced driver-assistance systems (ADASs) and electro-mobility solutions for electric vehicles and hybrid vehicles, growing market for portable devices such as wearables, and rising demand for multicore processors in military applications.

Unlike general-purpose computing systems, embedded systems are combination of hardware and software, either in a fixed or programmable capability, that are embedded within larger mechanical or electrical systems and are designed to perform specific functions, as shown in Fig. 4.1. Embedded systems are widely associated with microprocessors or microcontrollers, although some embedded systems can contain other technologies like digital signal processors (DSPs), complex programmable logic devices (CPLDs), application-specific integrated circuits (ASICs) and field programmable gate arrays (FPGAs).

L. L. Woo (✉)
Malaysian Space Agency (MYSA), National Space Centre, Banting, Malaysia
e-mail: elena@mysa.gov.my



B. Halak (ed.), *Hardware Supply Chain Security*,
https://doi.org/10.1007/978-3-030-62707-2_4

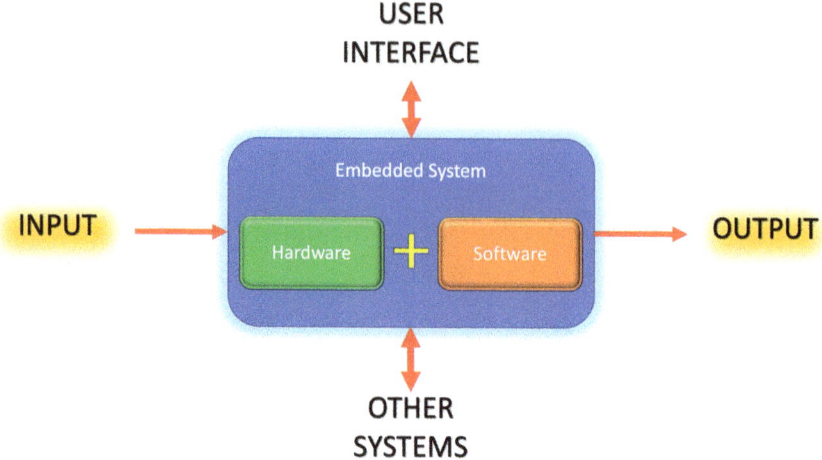


**Fig. 4.1** Embedded system

The hardware in an embedded system includes central processing unit like microcontrollers or microprocessors, power supply kit, memory devices, timers, communication ports and others. The software in an embedded system is a set of instructions to perform specific tasks. These sets of instructions are written using programming languages such as embedded C or C or even using programming software such as LabVIEW or Proteus.

## 4.2 Chapter Overview

This chapter is organised as follows. Section 4.3 discusses how embedded systems are classified either based on performance of microcontrollers or based on functional and performance requirements, while Sect. 4.4 discusses the characteristics of embedded systems. Section 4.5 discusses how anomalies found in a system may cause some high-level behaviour in the system to be anomalous, which may lead to an accidental situation. Section 4.6 presents the various type of threats that cause the system to behave anomalously, while Sect. 4.7 concludes the whole chapter.

## 4.3 Embedded Systems for Various Applications

As shown in Fig. 4.2, embedded systems can generally be classified into two categories: (a) based on performance of microcontrollers and (b) based on functional and performance requirements.

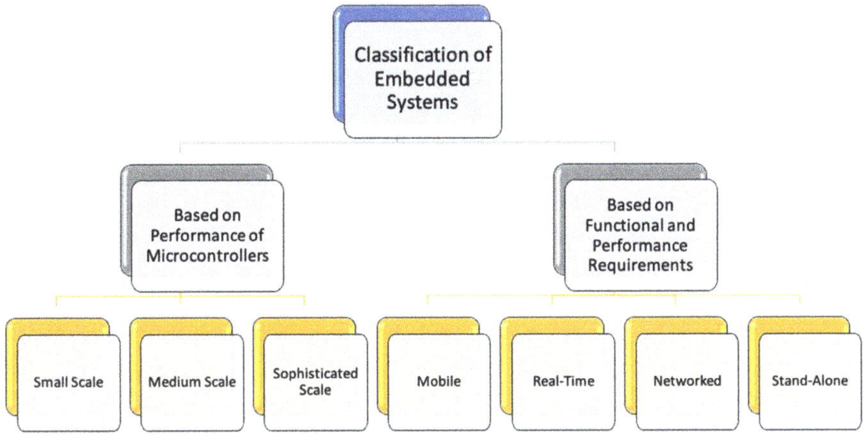


**Fig. 4.2** Classification of embedded system

1. **Microcontroller Performance Based Embedded Systems**
   Microcontroller performance based embedded systems can be divided into three types:
   - **Small-Scale**: Small-scale embedded systems are built for simple applications driven by low cost and low performance. It usually has the least complexity in terms of hardware and software and uses a single 8-bit or 16-bit microcontroller, which can be activated using a battery. An example of a small-scale embedded system is a digital watch.
   - **Medium-Scale**: Medium-scale embedded systems are slightly more complex in both hardware and software. It uses a single or few 16-bit or 32-bit microcontrollers, or Digital Signal Processors (DSPs) or Reduced Instructions Set Computers (RISCs). Some examples of medium-scale embedded systems are an Ipod or an MP3 player.
   - **Sophisticated-Scale**: Sophisticated embedded systems, typically used for cutting-edge applications, use more than 32-bit microcontroller. It has enormous complexity in both hardware and software, hence the need for reconfigurable or scalable processors and programming logic arrays. The system has limitations, such as the processing speed available in the hardware units. An example of a sophisticated embedded software is the on-board computer in a satellite.
2. **Functional and Performance Requirements Based Embedded System**
   Functional and performance requirements based embedded systems can further be divided into four types:
   - **Mobile**: Mobile embedded systems are used in portable embedded devices such as cameras, mobiles, cell phones, personal digital assistants (PDAs),

MP3 players and others. Mobile embedded systems are usually limited in memory and other resources.

- **Real-Time**: Real-time embedded systems are systems that require tasks to be performed in a particular time. In other words, these type of embedded systems follow timing constraints for completion of a task.
- **Networked**: Networked embedded systems are systems that are connected in a network to access resources. A connected network can be a Local Area Network (LAN), a Wide Area Network (WAN) or even the internet, with the connection type to be either wired or wireless. Majority of existing embedded system applications are of this type. One example of a network embedded system is a cleanroom building management system, where all the particles, humidity and temperature sensors are connected to a server and run on TCP/IP protocol.
- **Stand-alone**: As the name implies, a stand-alone system works on its own, with no requirement for a host system. It takes the input from the input ports, either digital or analog, processes, calculates and converts the data and gives the results through the connected device, such as a display. An example of a stand-alone system is a microwave oven or a water heater system.

Figure 4.3 shows the various applications for embedded systems, which range from household appliances to medical equipment, from consumer electronics to aerospace and defence, from telecommunication systems to automotive and transportation industries as well as safety and security systems.

**Fig. 4.3** Embedded system applications

- **Household Appliances**: Most of our household appliances, which we normally call electronics products, are in fact, embedded systems. Water heaters, microwave ovens, washing machines, refrigerators and dishwashers use embedded systems to provide the required features, efficiency and flexibility to users.
- **Medical Equipment**: Medical imaging such as Magnetic Resonance Imaging (MRI), Computed Tomography (CT), Single-Photon Emission Computed Tomography (SPECT) or Positron Emission Tomography (PET) for non-invasive internal investigations, embedded systems for vital signs monitoring or electronic stethoscope for sound amplification are just some of the applications developed using embedded systems.
- **Consumer Electronics**: Consumer electronics are one of the most common applications using embedded systems. Consumer electronics include devices used for entertainment (DVD players, flatscreen televisions, video games like PlayStation or XBox, remote control cars, etc.), devices for communications ( laptops, tablets, cell phones, etc.) and home–office activities (desktop computers, printers, scanners, paper shredders, etc.).
- **Aerospace and Defence**: Embedded systems for aerospace and defence applications have different requirements compared to the ones used for civilians. These systems must be highly secure to prevent data interception or reverse engineering, ruggedly designed to withstand harsh conditions in space or in the battlefield, and the source of the components used must be from a trusted entity to prevent unauthorised software from being loaded onto the devices. Satellites, surveillance and reconnaissance Unmanned Aerial Vehicle (UAV) and intelligence gathering operations are just some of the examples of embedded systems for aerospace and defence applications.
- **Telecommunication Systems**: Mobile computing users use a variety of embedded systems from telephone switches for the network to cell phones at the end user, while network computing users use dedicated routers and network bridges to route data for computers. Wireless communication uses transmitters and receivers of an antenna for data transfer.
- **Automotive and Transportation Industries**: Automotive and transportation industries are another area that utilises embedded systems heavily. Inertial guidance systems in the airplanes, motor or cruise control system, engine or body safety system, robotics in the assembly line and entertainment and multimedia in the car are just some of the embedded systems for this industry.
- **Safety and Security Systems**: Last but not least, embedded systems have contributed to the rise of safety and security systems. Sensors can be installed in our homes or offices, and complete visual feeds can be viewed on our mobile regardless of where we are in world.

## 4.4 Characteristics of Embedded Systems

Despite various types of applications and implementation methods that are available, embedded systems are bound by some common characteristics as shown in Fig. 4.4, which are briefly explained as follows:

- **Application Specific**: An embedded system is used to perform specific tasks related to its specific application, in contrast to a general-purpose computing system that executes a variety of applications [12]. This specific task is repeated continuously over the lifetime of the embedded system. For example, an embedded system designed for a washing machine will function only to wash clothes.
- **Limited Resources**: Due to the many reasons such as the nature of application, production costs and available hardware technology, embedded systems have tight constraints and limited resources concerning hardware resource, processor speed, power consumption and memory size [28]. For example, a microprocessor used in a general computing system operates at a clock speed above 2GHz, while the clock speed for a microcontroller varies between 20MHz and 300MHz, a fraction compared to the clock speed of a microprocessor.
- **Real Time**: Embedded systems have to perform tasks or interact with the external environment within specific timing constraints, where the correctness of the system depends on the output results as well as the time the results are produced [2]. For example, a car's brake system, if exceeds the time limit, may cause accidents.

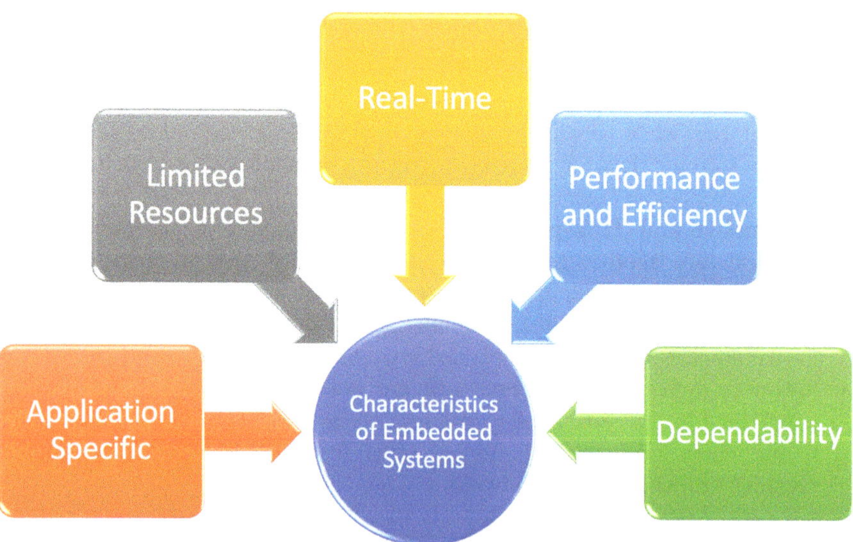


**Fig. 4.4** Characteristics of an embedded system

- **Performance and Efficiency**: Embedded systems are expected to achieve high performance, usually defined by the amount of tasks completed within certain execution time. Given the limited resources faced by these systems, embedded systems also have to be efficient in utilising the power consumption, memory utilisation and hardware resources [25].
- **Dependability**: Dependability is the ability to avoid service failures that are more frequent and more severe than what is acceptable [3, 11]. The common issues that arise in creating a dependable system are reliability, safety, security, availability, integrity and repairability [3, 11, 12, 25].
  - *Safety* means that the system is able to function without catastrophic failure or reduce the frequency of failures.
  - *Reliability* means ensuring that the system completes the task without experiencing any failure.
  - *Security* means the ability of the system to protect itself against deliberate or accidental intrusions.
  - *Availability* means that the system is able to deliver the service when it is required.
  - *Integrity* means that the system is protected against improper or unauthorised system alterations.
  - *Repairability* means that the system can undergo modifications or repairs.

## 4.5 Anomalies: Cause of Anomalous Behaviour

Figure 4.5 shows how the activated chain of threats of fault–error–failure may lead to an accidental situation, which in turn risks the human life. The failures occur when a fault has manifested itself an error and causes some high-level behaviour to be anomalous. Anomalous behaviour, or in short, anomalies, is a behaviour that does not conform to a normal, expected pattern and can also be identified as outliers, exceptions, peculiarities, contaminants or other terms according to the domain being studied [7, 18]. Grubbs [15] has defined an outlier as

> An outlying observation, or outlier, is one that appears to deviate markedly from other members of the sample in which it occurs.

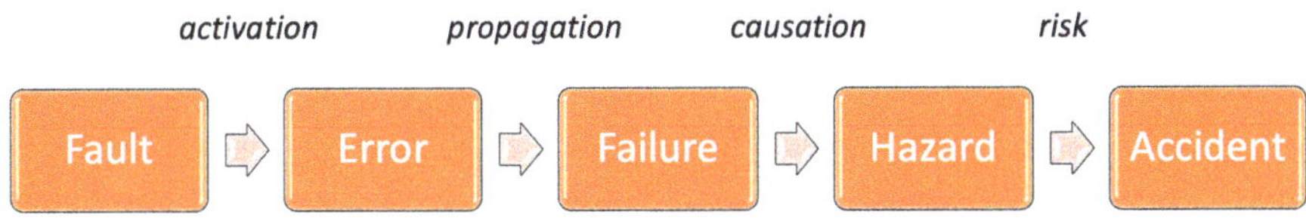


**Fig. 4.5** The fundamental chain of threat in a system [20]

In data mining, anomaly detection refers to scientific techniques applied to identify behaviours, events or data points that do not belong to the rest of the data in a dataset. Chandola [7] defined anomaly detection as

> The problem of finding patterns in data that do not conform to expected behaviour.

### 4.5.1 Type of Anomalies

Chandola [7] has also categorised anomalies into three different structures, namely (a) point anomalies, (b) contextual anomalies and (c) collective anomalies.

- **Point anomalies**:
  An individual data instance is deemed to be anomalous with respect to the rest of the data if it is "too far" from the rest of the data. Figure 4.6 shows an example of point anomalies where points O1 and O2 are deemed anomalous with respect to S1 and S2 as these points are located "too far" from the rest of the data.
- **Contextual anomalies**:
  An individual data instance is deemed to be anomalous in a specific context, but not otherwise. This type of anomaly is common in time-series data. Figure 4.7 shows an example of a contextual anomaly, T2. The graph shows the average monthly temperature collected at Southampton from January 2017

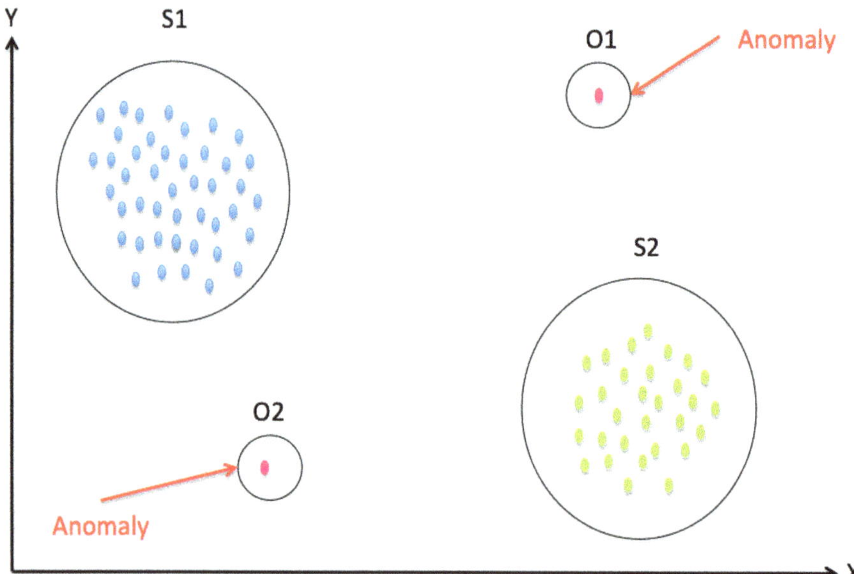


**Fig. 4.6** Point O1 and point O2 are point anomalies as they deviate far from the rest of the data

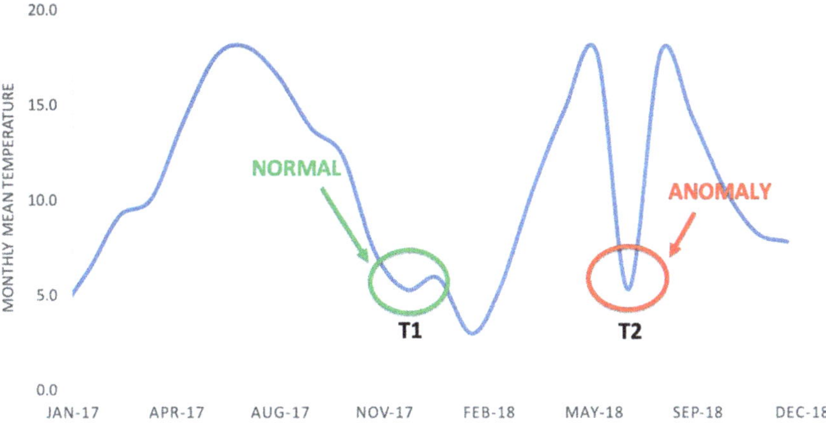


**Fig. 4.7** An example of contextual anomaly, $T2$, in a monthly temperature time series. Note that the temperature at $T2$ is similar to $T1$, but because it occurs in the month of July instead of November, hence it is considered anomalous

until December 2018 obtained from Southampton Weather Station.[1] The value of T2 bears similarity to the value of T1, but the low temperature of T2 happened during the summer period instead of the winter period, and hence it is considered anomalous.

- **Collective anomalies**:
  Individual data instances may not be anomalies themselves, but their occurrence together as a collection with respect to the rest of the data is deemed anomalous. Figure 4.8 shows an example of a collective anomaly corresponding to an Atrial Premature Contraction in a human electrocardiogram output. The value −5.5 as shown in Fig. 4.8 is not a value that deviates far from the rest of the data, but because it occurs consecutively for a period, hence it is considered anomalous.

## 4.6 Threats That Cause Anomalous Behaviour

The rise of embedded systems has also seen the rise of security threats in proportional rate [30]. As more and more embedded systems are integrated into our daily lives in a pervasive and "invisible" way, security becomes an important and critical aspect for developers in developing dependable embedded systems [27]. However, not all threats cause the system to behave anomalously. In this section, emphasis is given to threats that cause anomalous behaviour in the system. These threats can be grouped into several different categories as shown in Fig. 4.9 and are explained in the following sections.

[1]https://www.southamptonweather.co.uk.

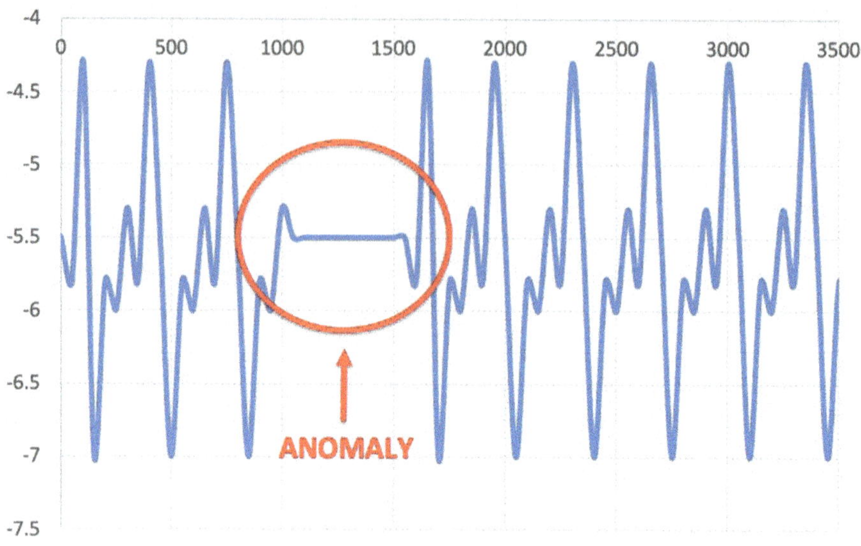


**Fig. 4.8** An example of collective anomaly corresponding to an Atrial Premature Contraction in a human electrocardiogram output

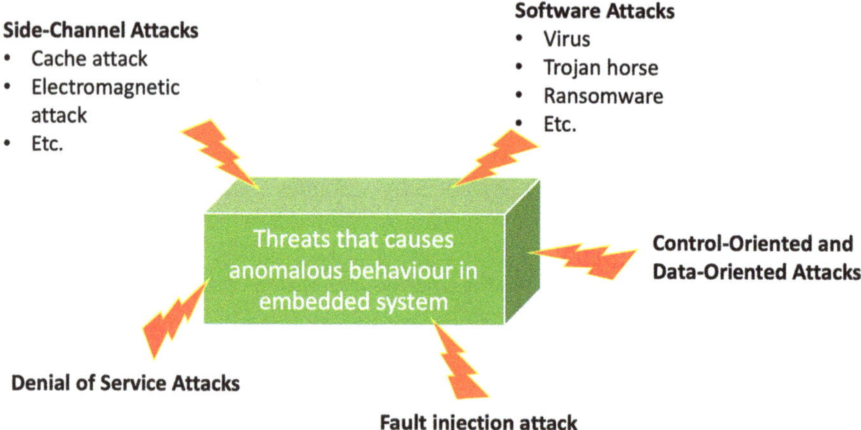


**Fig. 4.9** Threats in embedded system

## 4.6.1 Side-Channel Attacks

Side-channel attacks are non-invasive attacks that aim at extracting secrets from a chip or a system by monitoring, measuring and analysing the physical parameters. The physical parameters can be the supply current, power consumption, timing information, electromagnetic emission and others. Side-channel attacks pose serious threats especially to embedded systems that include cryptographic modules as many attacks have proven successful in breaking cryptographic algorithm and extracting the secret key to steal sensitive information. In side-channel attacks, attackers create an anomalous behaviour when stealing information from the victim. Side-channel attacks can be further divided into the following classes such as:

- **Cache attack**: Cache attacks are attacks based on the ability of the attacker to monitor cache accesses performed by the victim in a shared physical system such as in a virtual environment or in a cloud service [4, 22]. Inspired by the fact that intentional manipulation of cache usually causes anomalous behaviour, such as high cache misses, there have been numerous detection methods that focus on detecting cache attacks by focusing on some hardware events generated by the programs through hardware performance counters [6, 36, 37].
- **Power-monitoring attack**: Power-monitoring attacks or power-analysis attacks rely on the fact that the energy consumed by a hardware module depends on the switching activity of its transistors. Hence, by measuring the power consumption of the encryption device, and performing either Simple Power Analysis (SPA) or Differential Power Analysis (DPA), traces in the hardware module are obtained to decipher the secret key [10]. As shown in [26], detection of power-monitoring attack can be done by observing the instantaneous power-monitoring reading and power-monitoring amplitude using the ring oscillator-based power monitor and determined if either one of the values exceeds the reference value. If either one of the value exceeds the reference value set, an anomalous behaviour has occurred due to the attack.
- **Electromagnetic attack**: It is an attack based on leaked electromagnetic radiation, which can lead to sensitive data or information in the device [1]. *Van Eck phreaking* is a form of eavesdropping where special equipment are used to monitor the electromagnetic fields produced by the data in the device that represents the hidden information. These signals are then recreated with the purpose to spy on the device. Passive electromagnetic radiation is hard to detect as the system is not showing any sign of anomalous behaviour. However, authors in [35] have shown that for an active electromagnetic interference (EMI) where an attacker can use EMI to remotely, using easily available radio equipment, inject an attacking signal into the sensor system, the anomaly can be detected. Their work is based on the idea that if the sensor is turned off, the signal read by the microcontroller should be 0 V (or some other known value). If the microcontroller detects fluctuations in the sensor output, the attacking signal can be detected.

### *4.6.2 Software Attacks*

Software attacks refer to malicious programs written specifically to access someone's system in an unauthorised way or to deliberately vandalise another user's system. Malicious programs, or popularly known as malware, are harmful software, intentionally designed to wreck damage to a system, device, server, client or a network of computers. The presence of a malware in the computer can wreck all sorts of havoc, such as taking control of the machine, or monitoring user's actions and keystrokes, or silently sending all sorts of confidential data from user's computer or network to the attacker's home base. Software attacks can come in various forms. Some of the common types are:

- **Virus**: A virus is defined as a program that reproduces its own code by attaching itself to other executable files in such a way that the virus code is executed when the infected executable file is clicked. A virus requires a host to replicate, and it replicates itself every time the host is used and often focuses on destroying or corrupting the data.
- **Trojan horse**: A Trojan horse is a form of malware disguised as being a useful application or legitimate software. A Trojan horse usually contains a destructive code that will be activated when the application or software is used. While it does not replicate itself, once infected, Trojan horses often do something nasty and destructive to the system, such as allowing the computer to be spied or remotely controlled from the network.
- **Worm**: Another form of malware is called worm. Worm is a type of malware that can replicate itself and perform a variety of functions such as deleting files or sending documents via emails. It has the ability to infect other systems over networks and seriously impact the network traffic.
- **Ransomware**: Ransom malware, or ransomware in short, is a form of malware that prevents users from accessing their systems or files by encrypting it and demanding ransom payment in order to regain access. One common way for a computer to be inflected with a ransomware is through unsolicited email that might include malicious attachments disguised in a form of PDF or Word documents or links to malicious websites.
- **Spyware**: Spyware is a type of malicious software that is installed on a computing device without the user's knowledge. It aims to steal sensitive information such as account passwords or credit card numbers or track internet usage data and relay those information to external marketers.
- **Buffer overflow attack**: Buffer is a temporary storage place for data. When there is more data to be stored than what the buffer can originally store, the extra data overflow into adjacent memory space, corrupting or overwriting the data held in that space. In a buffer overflow attack, it allows the attacker to run arbitrary code or manipulate the coding errors to prompt malicious actions, such as the actions could trigger a response that damages files, changes data or unveils private information.
- **Backdoor**: A backdoor refers to any method by which authorised and unauthorised users are able to bypass normal authentication or encryption in a computer, product or embedded device and gain high-level user access on a system.

Anomaly detection on software attacks can be grouped into two: (a)anomaly-based intrusion detection and (b) fraud detection. Intrusion detection refers to detection of malicious activity such as break-ins, penetrations and other forms of computer abuse in a computer-related system. An anomaly-based intrusion detection system detects network or computer intrusions or misuse by monitoring its activity and classifying it as either normal or anomalous. The classification is based on rules or heuristics where if the current activity does not follow normal system operation and does not satisfy the rules that have been set, it will be classified as anomalous. As opposed to signature-based intrusion detection that has provided good detection rate, an anomaly-based intrusion detection system often has high false positives rate, where an event is wrongly classified as an attack. However, it is able to detect new or unfamiliar intrusion compared to signature-based intrusion detection system that is only capable of detecting existing or well-known attacks. [13] presented the various algorithms that have been used for anomaly-based intrusion detection systems.

Fraud detection refers to detection of criminal activities occurring in commercial organisations such as banks, credit card companies, insurance agencies, cell phone companies, stock market and so on, where malicious users obtain information from the organisations with the intention to use it in an unauthorised way. There are two approaches for fraud detection: (a) rule-based approach or (b) machine-learning approach. In a rule-based approach, fraudulent activities are detected by looking at evident or on-surface activities using fraud detection algorithms manually written by fraud analysts. However, rule-based approach is only able to detect obvious fraudulent activity, not to mention it requires extensive manual work to enumerate all possible scenarios with long processing time and involves multiple verification steps, which may not be a pleasant experience for users. In contrast, for fraud detection that uses machine-learning approach, it allows the creation of algorithms to automatically find hidden correlations between users' behaviour and possible fraud. This approach allows processing to be done in real time and reduces the multiple verification steps required in rule-based approach. In [29], the author presents three different unsupervised machine learning methods, such as autoencoder, Support Vector Machine (SVM) and Mahalanobis, to detect credit card fraud.

### 4.6.3 Denial-of-Service (DoS) Attack

A denial-of-service (DoS) attack occurs when the attacker aims to deny legitimate users the access to information systems, devices or other network resources by flooding the targeted host or network with traffic until the target host cannot respond or simply crashes, resulting in denial of service to legitimate users. DoS attacks can cost an organisation both time and money when their resources and services become inaccessible. A DoS attack is characterised by using a single computer to launch the attack, while a distributed denial-of-service (DDoS) attack uses multiple, distributed sources to launch the attack. In [16], the authors describe using machine learning algorithms to detect and protect the system from DoS attacks and other various attacks. A total of five different algorithms were used, such as Logistic Regressions (LR), SVM, decision tree, random forest and Artificial Neural Network (ANN).

### 4.6.4 Control-Oriented and Data-Oriented Attacks

Control-oriented attacks such as control-flow hijacking are attacks that exploit memory corruption vulnerability to divert the normal control flow of the program running on an embedded device to allow the execution of malicious code injected by the attacker or useful code sequences already in the program. When the victim's program control flow is being hijacked, it causes the program to perform operations that greatly deviate from its normal behaviour. As a consequence, the execution of regular and anomalous behaviour captured by hardware events is highly distinguishable. Previous researches have demonstrated that hardware-level information can be used for detecting control-oriented attacks with high accuracy [8, 34]. Data-oriented attacks according to [9] manipulate program's internal data variables without violating its control-flow integrity (CFI). Examples of data-oriented attacks are non-control data attacks, control-flow bending and data-oriented programming. According to [31], a successful detection scheme for this type of attacks requires the ability to identify when seemingly normal program code is an improper execution of the program. This necessitates monitoring a program's execution at much finer granularity. Data-oriented attacks uphold a valid control-flow during execution; hence, the anomalous behaviour due to this type of attacks is much harder to detect compared to control-oriented attacks. Therefore, to overcome this limitation, [31] proposed to incorporate both contextual and temporal information into a time series to increase the ability to discern anomalous behaviour from a normal behaviour.

### 4.6.5 Fault Injection Attacks

Fault injection attacks in the field of cryptography are attacks introduced with the attempt to steal information by introducing faults into the system's computation [5, 32]. In a fault injection attack, an attacker injects an intentional fault in a circuit by inducing a physical disturbance to the circuit's operation through the circuit's power supply or clock connections or by applying electromagnetic or laser pulses to the circuit and analyses the response of that circuit to the fault. The injected fault can cause disruption to the control flow or induce small differentials in cryptographic computations, which can then be used for cryptanalysis of the algorithm under attack. This fault injection for cryptanalysis is also known as Differential Fault Analysis (DFA). The aim of the attacker is to be able to inject an intentional fault, using a series of techniques to manipulate the environmental conditions of a circuit, that results in the desired fault effect such as stuck-at, set/reset or random fault [14] to extract cryptographic key material, to weaken cryptographic strength or to disable the security. Another type of fault injection attack is fault-sensitivity analysis (FSA), which uses the correlation between the data processed by the device and the sensitivity of the device to the fault injected. FSA involves repeating the cryptographic operation while, at the same time, degrades an operational parameter of the device gracefully until the circuit starts failing [33]. Countermeasures for fault injection attacks can be divided into either hardware based or software based. For example, researchers in [17] used sensors to detect any physical stress that could potentially result in fault injection. Software-based fault detection usually involves non-linear codes such as in [21] or redundancy such as in [19].

## 4.7 Conclusion

Compared to normal computing systems, embedded systems are more vulnerable to a range of attacks that aim to steal information, cripple the security system, hijacking the system or even destroy the system. This chapter provided an overview of embedded systems and discussed how embedded systems are classified either based on performance of microcontrollers or based on functional and performance requirements. This chapter also looked at common characteristics of embedded systems despite their various types of applications and implementation methods. Discussion on various type of anomalies and how anomalies in an embedded system can cause a system to malfunction was also presented in this chapter. This chapter concluded by identifying the different threats that cause the system to behave anomalously.

## References


1. D. Agrawal, B. Archambeault, J.R. Rao, P. Rohatgi, The EM side—channel(s), in *Cryptographic Hardware and Embedded Systems, CHES*, pp. 29–45 (Springer, Berlin, Heidelberg, 2003)
2. A. Armoush, Design patterns for safety-critical embedded systems. Ph.D. thesis, RWTH Aachen University (2010)
3. A. Avizienis, J.C. Laprie, B. Randell, C. Landwehr, Basic concepts and taxonomy of dependable and secure computing. IEEE Trans. Dependable Secure Comput. **1**(1), 11–33 (2004). https://doi.org/10.1109/TDSC.2004.2
4. M.M. Bazm, T. Sautereau, M. Lacoste, M. Sudholt, J.M. Menaud, Cache-based side-channel attacks detection through Intel cache monitoring technology and hardware performance counters, in *2018 Third International Conference on Fog and Mobile Edge Computing (FMEC)*, pp. 7–12 (2018)
5. J. Breier, X. Hou, Y. Liu, Fault attacks made easy: Differential fault analysis automation on assembly code. IACR Trans. Cryptogr. Hardware Embedded Syst. **2018**(2), 96–122 (2018). https://doi.org/10.13154/tches.v2018.i2.96-122
6. S. Briongos, G. Irazoqui, P. Malag6n, T. Eisenbarth, Cacheshield: Protecting legacy processes against cache attacks. CoRR ArXiv:abs/1709.01795 (2017)

7. V. Chandola, A. Banerjee, V. Kumar, Anomaly detection: A survey. ACM Comput. Surv. **41**(3), 15:1–15:58 (2009). https://doi.org/10.1145/1541880.1541882
8. S. Chen, J. Xu, E.C. Sezer, P. Gauriar, R.K. Iyer, Non-control-data attacks are realistic threats, in *Proceedings of the 14th Conference on USENIX Security Symposium, SSYM'05*, vol. 14 (USENIX Association, USA, 2005), p. 12
9. L. Cheng, K. Tian, D. Yao, L. Sha, R.A. Beyah, Checking is believing: Event-aware program anomaly detection in cyber-physical systems. IEEE Trans. Dependable Secure Comput., 1–1 (2019)
10. D. Das, S. Maity, S.B. Nasir, S. Ghosh, A. Raychowdhury, S. Sen, High efficiency power side-channel attack immunity using noise injection in attenuated signature domain, in *2017 IEEE International Symposium on Hardware Oriented Security and Trust (HOST)* (2017). https://doi.org/10.1109/hst.2017.7951799.
11. E. Dubrova, *Fault-Tolerant Design* (Springer, 2013)
12. M.D.P. Emilio, *Features of Embedded Systems* (Springer International Publishing, Switzerland, 2015), pp. 25–31. https://doi.org/10.1007/978-3-319-06865-7_2
13. P. Garcí-Teodoro, J. Dí-Verdejo, G. Maciá-Fernández, E. Vázquez, Anomaly-based network intrusion detection: Techniques, systems and challenges. Comput. Secur. **28**(1), 18–28 (2009). https://doi.org/10.1016/j.cose.2008.08.003. https://www.sciencedirect.com/science/article/pii/S0167404808000692
14. N.F. Ghalaty, Fault attacks on cryptosystems: Novel threat models, countermeasures and evaluation metrics. Ph.D. thesis, Virginia Polytechnic Institute and State University (2016)
15. F.E. Grubbs, Procedures for detecting outlying observations in samples. Tech. Rep. 1, American Statistical Association and American Society for Quality (1969). https://www.jstor.org/stable/1266761
16. M. Hasan, M.M. Islam, M.I. Islam, M. Hashem, Attack and anomaly detection in IoT sensors in IoT sites using machine learning approaches. Internet of Things **7**, 100059 (2019). https://doi.org/10.1016/j.iot.2019.100059
17. W. He, B. Jakub, S. Bhasin, Cheap and cheerful: A low-cost digital sensor for detecting laser fault injection attacks, in *6th International Conference on Security, Privacy and Applied Cryptography Engineering (SPACE)*, pp. 27–46 (2016). https://doi.org/10.1007/978-3-319-49445-6_2
18. V.J. Hodge, J. Austin, A survey of outlier detection methodologies. Artif. Intell. Rev. **22**(2), 85–126 (2004)
19. M. Joye, P. Manet, J. Rigaud, Strengthening hardware AES implementations against fault attacks. IET Inf. Secur. **1**(3), 106–110 (2007)
20. P. Koopman, Reliability, safety and security in everyday embedded systems. Dependable Computing, Lecture Notes in Computer Science, vol. 4746/2007 (2007). https://doi.org/10.1007/978-3-540-75294-3_1
21. K.J. Kulikowski, M.G. Karpovsky, A. Taubin, Fault attack resistant cryptographic hardware with uniform error detection. *Fault Diagnosis and Tolerance in Cryptography* (Springer, Berlin, Heidelberg, 2006), pp. 185–195
22. F. Liu, H. Wu, K. Mai, R.B. Lee, Newcache: Secure cache architecture thwarting cache side-channel attacks. IEEE Micro **36**(5), 8–16 (2016)
23. MarketsAndMarkets.com, Embedded systems market by hardware (MPU, MCU, application specific IC / application specific standard product, DSP, FPGA, and memory), software (middleware and operating system), system size, functionality, application, region - global forecast to 2025 (2020). https://www.marketsandmarkets.com/Market-Reports/embedded-system-market-98154672.html
24. P. Marwedel, *Embedded System Design* (Kluwer, 2003)
25. P. Marwedel, *Embedded System Design: Embedded Systems Foundations of Cyber-Physical Systems, and the Internet of Things* (Springer International Publishing, 2018)
26. A.L. Masle, W. Luk, Detecting power attacks on reconfigurable hardware, in *22nd International Conference on Field Programmable Logic and Applications (FPL)*, pp. 14–19 (2012)

27. D. Papp, Z. Ma, L. Buttyan, Embedded systems security: Threats, vulnerabilities, and attack taxonomy, in *2015 13th Annual Conference on Privacy, Security and Trust (PST)*, pp. 145–152 (2015)
28. H. Psaier, S. Dustdar, A survey on self-healing systems: Approaches and systems. Computing **91**(1), 43–73 (2011). https://doi.org/10.1007/s00607-010-0107-y
29. M. Rezapour, Anomaly detection using unsupervised methods: Credit card fraud case study. Int. J. Adv. Comput. Sci. Appl. **10**(11) (2019). https://doi.org/10.14569/IJACSA.2019.0101101
30. K. Siratla, B. Ankisetty, A.O. Chaitanya, Embedded system threats: Security threats and solutions. Int. J. Sci. Dev. Res. **3**, 118–121 (2018)
31. G. Torres, Z. Yang, Z. Blasingame, J. Bruska, C. Liu, Detecting non-control-flow hijacking attacks using contextual execution information, in *Proceedings of the 8th International Workshop on Hardware and Architectural Support for Security and Privacy, HASP 2019* (Association for Computing Machinery, New York, NY, USA, 2019). https://doi.org/10.1145/3337167.3337168
32. M. Tunstall, D. Mukhopadhyay, S. Ali, Differential fault analysis of the advanced encryption standard using a single fault, in *Information Security Theory and Practice. Security and Privacy of Mobile Devices in Wireless Communication* (Springer, Berlin, Heidelberg, 2011), pp. 224–233
33. F. Valencia, I. Polian, F. Regazzoni, *Fault Sensitivity Analysis of Lattice-Based Post-Quantum Cryptographic Components* (Springer, 2019), pp. 107–123. https://doi.org/10.1007/978-3-030-27562-4_8
34. X. Wang, R. Karri, NumChecker: detecting kernel control-flow modifying rootkits by using hardware performance counters, in *Proceedings of the 50th Annual Design Automation Conference* (ACM, 2013), p. 7. https://doi.org/10.1145/2463209.2488831
35. Y. Zhang, K. Rasmussen, Detection of electromagnetic interference attacks on sensor systems, in *IEEE Symposium on Security and Privacy (S&P)* (2020)
36. T. Zhang, Y. Zhang, R. Lee, CloudRadar: A real-time side-channel attack detection system in clouds, in *International Symposium on Research in Attacks, Intrusions, and Defenses*, vol. 9854, pp. 118–140 (2016). https://doi.org/10.1007/978-3-319-45719-2_6
37. B. Zheng, J. Gu, C. Weng, CBA-detector: An accurate detector against cache-based attacks using HPCs and pintools, in *Advanced Parallel Processing Technologies* (Springer International Publishing, 2019), pp. 109–122

# Chapter 5
# Hardware Performance Counters (HPCs) for Anomaly Detection

**Lai Leng Woo**

## 5.1 Introduction

Most modern processors have special, on-chip hardware that can monitor performance known as *Hardware Performance Counters* or HPCs. HPCs are sets of special-purpose counters built into processors such as Intel Pentium, ARM, Cray, PowerPC, UltraSparc and MIPS architectures [29]. As shown in Fig. 5.1, these HPCs are special registers that are made available in the processor to track low-level Performance Monitoring Events (PMEs) within the processors, such as the number of cache misses, the number of instructions retired and the number of branch instructions retired in real time.

The type of events monitored using these counters can be categorised into two: (a) architectural and (b) non-architectural. Architectural events are events that remain consistent across different processor architectures such as instructions or branches. Non-architectural events comprise of events that are specific to the micro-architecture of a given processor, such as branch predictions or cache assesses. Different processor architectures may have different non-architectural events. Not only that, processors that undergo enhancements will also have different non-architectural events [8]. Data collected using this special hardware provide performance information on applications, the operating system, as well as the processor. These data can be used to guide performance improvement efforts by providing information that helps programmers to tune the algorithms used by the applications and operating system, as well as the code sequences implementing those algorithms. Besides for performance improvement efforts, these data can also be used to detect anomalies or malicious activities that might occur in a system.

L. L. Woo (✉)
Malaysian Space Agency (MYSA), National Space Centre, Banting, Malaysia
e-mail: elena@mysa.gov.my



B. Halak (ed.), *Hardware Supply Chain Security*,
https://doi.org/10.1007/978-3-030-62707-2_5

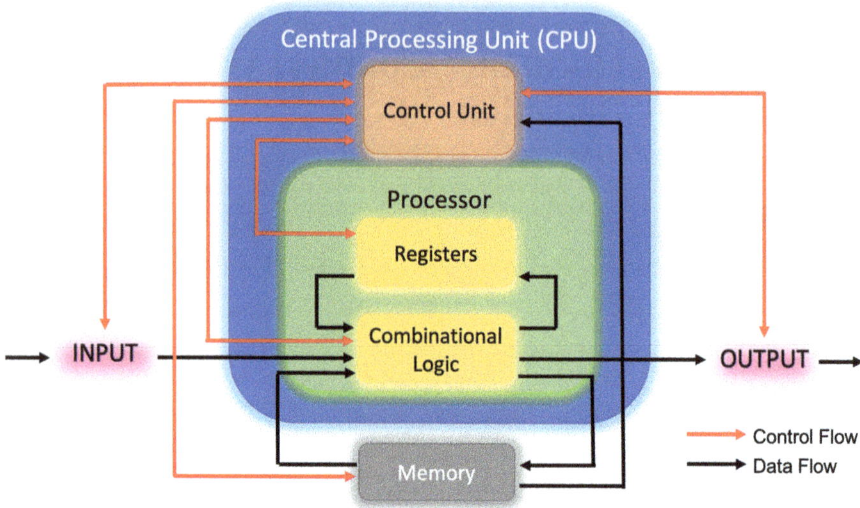


**Fig. 5.1** A block diagram of a basic computer with a single processor

## 5.2 Chapter Overview

Section 5.3 will provide a general view of the type and characteristics of the Hardware Performance Counters (HPCs) for some processor architectures such as Intel IA-32 architecture, ARM Cortex-M Series and PowerPC. Section 5.4 will discuss the operation principles of HPCs and what are the available tools that can be used to access these counters. Section 5.5 will present the overview of applications using HPCs. Two case studies where HPCs are used for anomaly detection are reviewed in Sects. 5.6 and 5.7. Finally, Sect. 5.8 concludes the chapter.

## 5.3 Type and Characteristics of Hardware Performance Counters (HPCs)

HPCs are part of the wider Performance Monitoring Unit (PMU) built into most modern processors. A PMU consists of two components: performance event select registers and event counters. A counter is paired with an event select register to monitor the occurrences of specific signals related to processor's function. These specific signals are known as Performance Monitoring Event (PME).

Different processor has a different terminology for the event select registers. For example, in an Intel x86 processor, the performance event select registers are known as model-specific registers (MSRs) [14], and for an ARM Cortex-A series processor, the registers are controlled through the event bus, PMUEVENT [2]. The PMU is interrupt based, such that an interrupt is generated after a certain interval of time or the number of occurrences of the desired event exceeds a predefined threshold. In other words, it is possible for PMU to do either time-based sampling or event-based sampling. The counters are incremented on an instruction-by-instruction basis, thus ensuring accurate results [17, 18]. As these counters are built in, there are no additional overheads to access the enormous information available in the CPU.

The number of available counters in each processor is limited and the available PMEs differ from one processor to another due to architectural differences. The number of available counters limits the number of PMEs that can be monitored in real time. For example, an Intel Atom has only two programmable performance monitoring counter registers per processor core. This means that only two PMEs can be monitored simultaneously. Therefore, it is not practical to utilise more microarchitectural events than the number of available counters to achieve high accuracy as it requires executing the application multiple times, since the hardware can only count a small subset of events concurrently [24]. It has been shown, however, that a single counter is sufficient to describe the behaviour of a program [3, 6]. Table 5.1 shows the number of available counters and the number of available PMEs for some common processors such as Intel, ARM, POWER4 and UltraSparc II. The total available counters in the processor and the number of available PMEs are taken from the technical reference manual of each processor.

Tables 5.2 and 5.3 list some PMEs that can be observed from an Intel® and ARM architecture.

**Table 5.1** Number of available counters and events for some processors

| Processor | Number of available HPCs | Number of available PMEs |
|---|---|---|
| Intel Atom [14] | 2 + 3 (fixed functions) | 129 |
| Intel Core i7 Nehalem [14] | 4 + 3 (fixed functions) | 129 |
| ARM Cortex-A9 [2] | 6 | 57 |
| POWER4 [4] | 8 | >100 |
| UltraSparc II [27] | 2 | >4 bil |

**Table 5.2** Pre-defined architectural performance monitoring events for Intel® architecture [14]

| Bit Position CPUID.AH.EBX | Performance Monitoring Event (PME) Name |
|---|---|
| 0 | UnHalted Core Cycles |
| 1 | Instruction Retired |
| 2 | UnHalted Reference Cycles |
| 3 | LLC Reference |
| 4 | LLC Misses |
| 5 | Branch Instruction Retired |
| 6 | Branch Misses Retired |

**Table 5.3** Examples of performance monitoring events for ARM architecture [2]

| Name | Event number | Description |
|---|---|---|
| PMUEVENT[0] | 0x00 | Software increment |
| PMUEVENT[1] | 0x01 | Instruction cache miss |
| PMUEVENT[2] | 0x02 | Instruction micro-TLB miss |
| PMUEVENT[3] | 0x03 | Data cache miss |
| PMUEVENT[4] | 0x04 | Data cache access |
| PMUEVENT[5] | 0x05 | Data read |
| PMUEVENT[6] | 0x06 | Data writes |
| ⋮ | ⋮ | ⋮ |
| PMUEVENT[56] | 0xA4 | PLE FIFO overflow |
| PMUEVENT[57] | 0xA5 | PLE request programmed |

## 5.4 Hardware Performance Counters (HPCs) Operation Principles

As discussed in Sect. 5.3, HPCs are implemented on a processor and have both an event select register and a cycle counter. In an Intel Pentium processor, such as Intel Pentium Pro, II and III, the event select register are 40-bit wide, while the cycle counter has a width of 64 bits. The select registers can be used to count a large number of performance events, such as instructions retired, cache accesses, misses, stalls in various components of the chip, etc. Each time the programmed event occurs, the cycle counter is incremented, irrespective of the process that has control of the CPU. The control register can be programmed to throw an interrupt when the counter register overflows, and this interrupt can be handled at software level. Further information regarding the implementation of HPCs on an Intel Pentium processor can be found in [14].

### *5.4.1 Performance Monitoring Tools*

There are several tools available to give users access to the hardware performance counters (HPCs). The most common tool available is Performance Application Programming Interface (or in short, PAPI) [5]. PAPI is a library tool that encapsulates low-level access to the hardware and provides an Application Programming Interface (API) for programmers to set up, start, stop and read the counters. PAPI abstracts some standard hardware events and provides a cross-platform standard naming for many useful events, such as cycle count, floating point instructions and others. Figure 5.2 shows the architecture of PAPI.

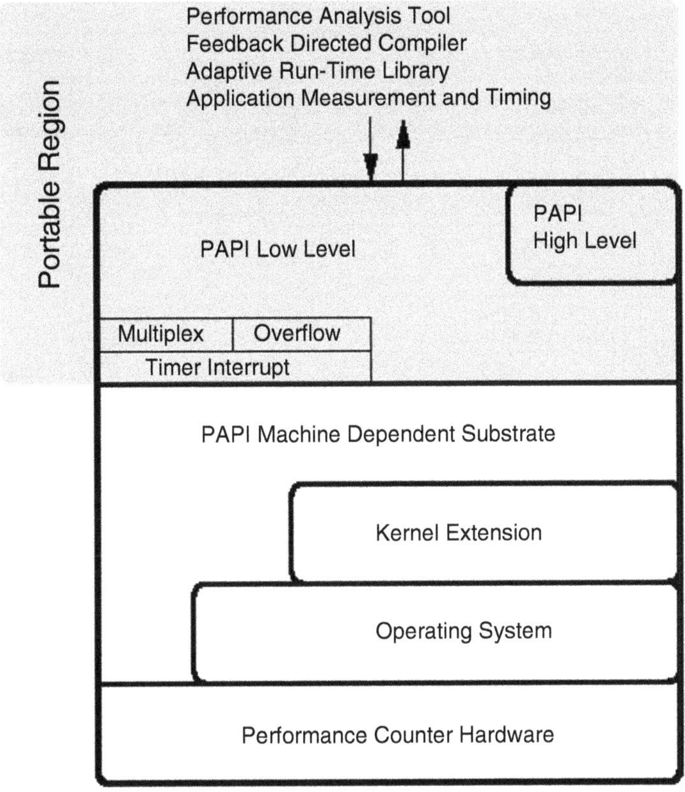


**Fig. 5.2** Architecture of Performance Application Programming Interface (PAPI)

Another example is PerfMon2 [20]. Perfmon2 is a standardised, generic interface to access the performance monitoring unit (PMU) of a processor. It is portable across all PMU models and architectures and supports system-wide and per-thread monitoring, counting and sampling. Linux's *perf*, which has also been called as Performance Counters for Linux (PCL), Linux perf events (LPE), or perf_events, is also another widely used tool for performance monitoring using hardware counters. *perf* is a kernel-based subsystem that provides a framework for collecting and analysing performance data. *perf* is also capable of lightweight profiling and can instrument CPU performance counters, tracepoints, kprobes and uprobes. Figure 5.3 shows the example of events that can be monitored using Linux's *perf* tool.

Performance Counter Monitor (Intel PCM) [33] as shown in Fig. 5.4 is another tool developed by Intel to estimate the internal resource utilisation of the latest Intel®Xeon®and Core™ processors and gain a significant performance boost. The package includes easy-to-use command line and graphical utilities. Intel PCM has

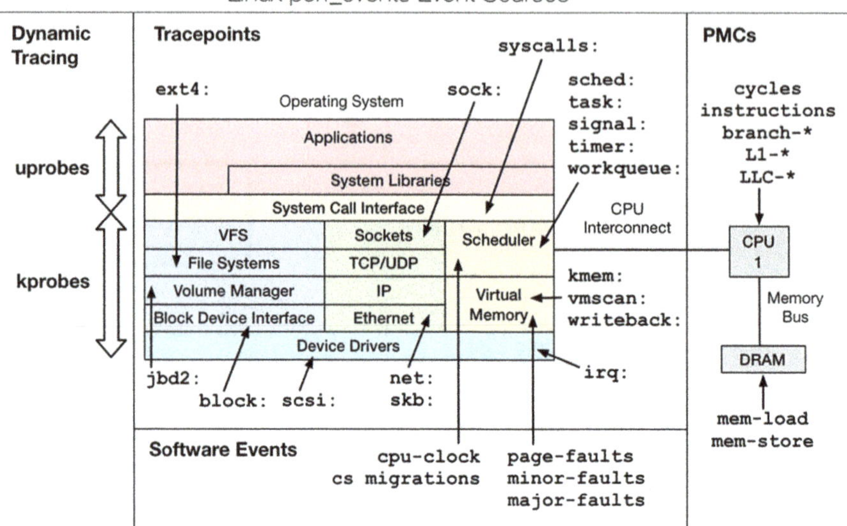


**Fig. 5.3** Example of events that can be monitored using Linux's *perf*

been discontinued since 2012 and is superseded by Processor Counter Monitor (PCM) [21]. PCM is an application programming interface (API) and a set of tools based on the API to monitor performance and energy metrics of Intel®Core™, Xeon®, Atom™ and Xeon Phi™ processors. PCM works on Linux, Windows, Mac OS X, FreeBSD and DragonFlyBSD operating systems and provides a number of command line utilities for real-time monitoring. Figure 5.5 displays the graphical front end of PCM, using pcm Grafana dashboard, one of PCM's many command line utilities.

There are also some commercial tools that provide access to the hardware counters through a graphical interface, such as Intel's VTune performance analyser [28]. VTune does not requires recompilation but samples the execution based on hardware or operating system events and combines the results with other analyses to provide tuning advice. Figure 5.6 shows the interface of the Intel's VTune Profiler. Other tools include Rabbit [13], OProfile [7], LiMiT [9] and TipTop [23].

## 5.5 Application of Hardware Performance Counters (HPCs)

HPCs were originally designed to be used as hardware verification or debugging tools for performance analysis or tuning purposes but have since been used for performance evaluation, workload estimation, detection of malicious activities and anomaly detection.

```
{C:\} - Far 2.0.1420 x86 Administrator
 IPC   : instructions per CPU cycle (0..4 on Nehalem and Westmere)
 FREQ  : relation to nominal CPU frequency=current CPU frequency/nominal frequency (includes Intel(r
) Turbo Boost Technology)
 L3MISS: L3 cache misses
 L2MISS: L2 cache misses (including other core's L2 cache *hits*)
 L3HIT : L3 cache hit ratio (0.00-1.00)
 L2HIT : L2 cache hit ratio (0.00-1.00)
 L3CLK : ratio of CPU cycles lost due to L3 cache misses (0.00-1.00), in some cases could be >1.0 du
e to a higher memory latency
 L2CLK : ratio of CPU cycles lost due to missing L2 cache but still hitting L3 cache (0.00-1.00)
 READ  : bytes read from memory controller (in GBytes)
 WRITE : bytes written to memory controller (in GBytes)

 Core (SKT) | IPC  | FREQ  | L3MISS | L2MISS | L3HIT | L2HIT | L3CLK | L2CLK  | READ  | WRITE

   0    0     0.09   0.60    551 K    569 K    0.03    0.04    2.10    0.02     N/A     N/A
   1    0     0.00   0.60      6      163      0.96    0.43    0.00    0.00     N/A     N/A
   2    0     0.34   0.60    691       42 K    0.98    0.10    0.01    0.12     N/A     N/A
   3    0     0.01   0.60     10     6134      1.00    0.01    0.00    0.02     N/A     N/A
   4    0     0.15   0.60     97 K    120 K    0.19    0.04    0.04    0.04     N/A     N/A
   5    0     0.01   0.60     23     6435      1.00    0.03    0.00    0.01     N/A     N/A
   6    0     0.06   0.60   4856       37 K    0.87    0.00    0.05    0.07     N/A     N/A
   7    0     0.01   0.60     55       10 K    0.99    0.00    0.00    0.02     N/A     N/A
   8    0     0.19   0.60     82 K    124 K    0.34    0.03    0.81    0.09     N/A     N/A
   9    0     0.01   0.60    462     1775      0.74    0.00    0.00    0.00     N/A     N/A
  10    0     0.01   0.60     17      953      0.98    0.00    0.00    0.00     N/A     N/A
  11    0     0.29   0.60     12 K     64 K    0.81    0.07    0.15    0.14     N/A     N/A
  12    1     0.05   0.60   1812       27 K    0.93    0.01    0.01    0.04     N/A     N/A
  13    1     0.01   0.60     55       11 K    1.00    0.00    0.00    0.02     N/A     N/A
  14    1     0.55   0.60   4491       48 K    0.91    0.25    0.02    0.04     N/A     N/A
  15    1     0.01   0.60     46       15 K    1.00    0.00    0.00    0.01     N/A     N/A
  16    1     0.02   0.60    271       22 K    0.99    0.00    0.00    0.02     N/A     N/A
  17    1     0.01   0.60     32       23 K    1.00    0.00    0.00    0.02     N/A     N/A
  18    1     0.03   0.60    809       35 K    0.98    0.02    0.00    0.03     N/A     N/A
  19    1     0.02   0.60    392       25 K    0.98    0.01    0.00    0.02     N/A     N/A
  20    1     0.03   0.60   2889       42 K    0.93    0.00    0.01    0.03     N/A     N/A
  21    1     0.01   0.60     87       20 K    1.00    0.00    0.00    0.02     N/A     N/A
  22    1     0.12   0.60     12 K     99 K    0.88    0.13    0.04    0.07     N/A     N/A
  23    1     0.01   0.60    142       23 K    0.99    0.00    0.00    0.02     N/A     N/A
-----------------------------------------------------------------------------------------------
 SKT    0     0.08   0.60    750 K    984 K    0.24    0.04    0.51    0.04    0.01    0.00
 SKT    1     0.08   0.60     23 K    396 K    0.94    0.08    0.01    0.03    0.00    0.00
-----------------------------------------------------------------------------------------------
 TOTAL  *     0.08   0.60    773 K   1385 K    0.44    0.05    0.19    0.03    0.01    0.00

 PHYSICAL CORE IPC: 0.16 => corresponds to 4.08 % core utilization

Intel(r) QPI traffic estimation in bytes (traffic coming to CPU/socket through QPI links):

               QPI0     QPI1
-----------------------------------------------------------------------------------------------
 SKT    0     1997 K   1989 K
 SKT    1     1760 K     13 K
-----------------------------------------------------------------------------------------------
Total QPI traffic: 5761 K     QPI traffic/Memory controller traffic: 0.34
^C
C:\>
1Left  2Right  3View..  4Edit..  5Print  6MkLink  7Find  8Histry  9Video  10Tree  11ViewHs  12FoldHs
```

Fig. 5.4 Intel Performance Counter Monitor (Intel PCM)

Fig. 5.5 Processor Counter Monitor (PCM)

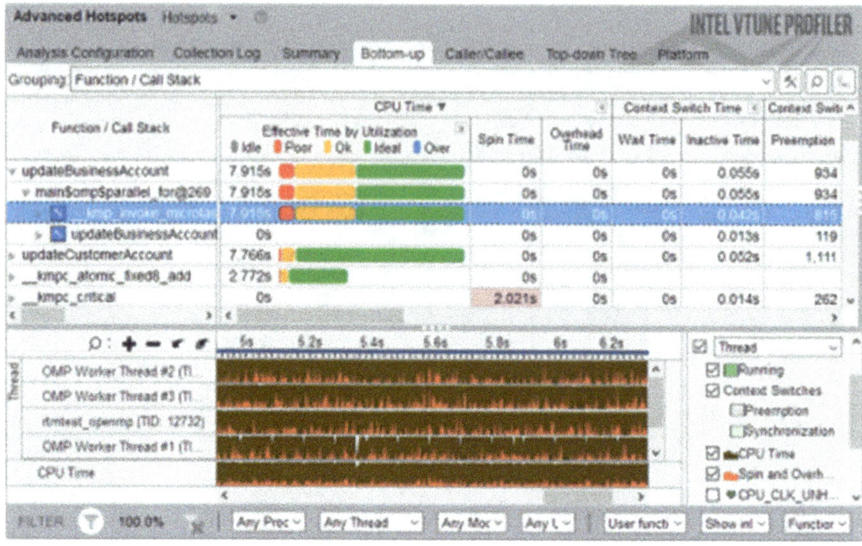


Fig. 5.6 Intel's VTune Profiler

### 5.5.1 Performance Evaluation

As the name implies, hardware performance counters provide data to describe the operating system and the application's performance on the processor. According to [26], hardware events for performance evaluation can be grouped into five categories: (a) program characterisation, (b) memory accesses, (c) pipeline stalls, (d) branch prediction and (e) resource utilisation, as depicted in Fig. 5.7.

- **Program characterisation**: Certain PMEs such as number of branch instructions, load instructions and store instructions are used to define the attributes or characteristics of an operating system or a program. These type of PMEs are architectural PMEs that are independent of processor implementation. Authors in [18] developed four different algorithms that utilise PMEs to reduce the estimation error, which saw an improvement of 40% in obtaining accurate data of all the multiplexed events for all the workloads for the purpose of performance measurement.
- **Memory accesses**: Events such as number of cache hits or misses are used to evaluate the performance analysis of the processor's memory hierarchy.
- **Pipeline stalls**: Pipeline stall PME helps users to analyse how well the program's instructions flow through the pipeline. This is especially useful to processors with deep pipelines to ensure the pipeline is filled with useful instructions. In order to determine if an application causes partial register stalls, user can configure the PMEs and event counters to count the number of partial register stall cycles as the application runs. If the frequency of these cycles as a percentage of the

**Fig. 5.7** Categories of hardware events for performance evaluation

application's total cycles is significant, finding and eliminating the partial register stalls could increase performance.

- **Branch prediction**: Branch prediction event is an indication of how well the processor can predict accurately the outcome of branches and keep correct instructions flowing into the processor's pipeline.
- **Resource utilisation**: Some PMEs can be used monitor the resource utilisation in a processor. For an example, counting the number of cycles of an integer or a floating-point divider can indicate whether the particular resource is utilised effectively or not. Wang et al. [30] used HPCs to monitor and quantify the interference between virtual machines located in the same host and competing for shared physical resources. Using Last-Level Cache (LLC) miss rates, one of the counters available, the data are fed into the interference prediction model to predict performance degradation between virtual machines and through the information gathered, it can determine which virtual machine is utilising most of the resources.

### 5.5.2 Workload Estimation

Workload estimation is an important step towards achieving good resource management. One area that requires good resource management is power consumption in both classical data centre workloads and high performance computing environments. Among the power-saving techniques used in current processors, one of it is Dynamic Voltage and Frequency Scaling (DVFS). One of the major attempts to increase energy efficiency of computing systems is by adapting the voltage and frequency of compute cores or other processor parts to the current system load. As the consumed power is proportional to the frequency and the squared voltage, a reduction in both will significantly reduce the processor's power consumption. Authors in [25] argued that using the load of the processor core to reduce the frequency of idling processor cores is not precise or sufficient enough and thus, proposed to use a more sophisticated method called *instructions per memory access* to make decision on the frequency. The benchmark results from two different x86 64 test systems showed that the average power consumption while running a real-life-like workload can be reduced by up to four percent or even more for very memory intensive applications. Although the average runtime increases slightly, the energy consumption, which is power consumption multiplied by runtime, can be reduced by up to two percent on average. Another example of how HPCs are used for workload estimation is shown in [22] where the authors proposed to monitor L1 cache activity counters in order to estimate the workload and set the Dynamic Voltage and Frequency Scaling (DVFS) based on the estimated workload. This method resulted in energy saving of 23% compared to the on-demand frequency setting policy used in Linux.

### 5.5.3 Detection of Malicious Activities

Detection of malicious activities using hardware performance counters is another research area actively pursued by various researchers. For example, in [15], the authors proposed BRAIN, which stands for BehaviouR based Adaptive Intrusion detection in Networks and that uses statistics gathered from hardware, network and application to detect and mitigate Distributed Denial of Service (DDoS) attacks on an application. The HPCs that form the hardware in BRAIN are used to characterise the host behaviour during load and attack. The result shows that by correlating the HPCs with network statistics and application statistics can successfully detect DDoS with high accuracy, low cost and performance overheads. In [31], the authors presented NumChecker, a Virtual Machine Monitor (VMM)-based framework that securely and efficiently monitors the execution of system calls to detect kernel rootkits by leveraging on existing HPCs. In [32], ConFirm is a low-cost technique that uses HPCs as a signature to verify the execution of the computational paths in order to detect malicious modification of firmware in embedded control systems.

Authors in [17] used HPCs for static and dynamic integrity checking of programs to detect malicious program modifications at load time (static) and at runtime (dynamic). By assuming there is a correlation between PMEs and program to be checked, the authors determined which PMEs to be used in the profile making, then built a mathematical model and used that model for runtime integrity checking.

### *5.5.4 Anomaly Detection*

The presence of anomalies in embedded systems are reported to be on the rise, and this leads to a rise in researchers proposing various methods for anomaly detecting. As highlighted in [16], a system that behaves normally exhibits a certain pattern, thus any behaviour that deviates from that normal pattern should be identifiable. In anomaly detection, it is important to understand what constitutes normal behaviour of a system, and what causes the system to behave anomalously. As the saying goes, there is NO "one-size-fits-all" solution for anomaly detection due to the nature of data and the type of anomalies that occur in different application domains. Some use statistical-based techniques [34], while others use clustering-based techniques [10], Support Vector Machine (SVM) [1], Hidden Markov Model (HMM) [19], Long Short-Term Memory (LSTM) [19] and many more.

## 5.6 Case Study 1: Anomaly Detection Using Support Vector Machine Method

In this section, we review one example of how a supervised anomaly detection can be done using hardware performance counters (HPCs) for monitoring. In this case study, Support Vector Machine (SVM), which is a supervised machine learning technique, is used as a classifier for model training based on features obtained from the HPCs. This first case study is based on the works of [1]. In this work, the authors aimed to detect attacks that cause application to behave anomalously by comparing the targeted application with the untarnished version of the application.

### *5.6.1 Methodology*

Figure 5.8 outlined the methodology used in the paper, which will be further explained below:

- **Benchmarks**:
  Using the CHStone benchmark suite from [12], the authors selected twelve benchmarks to be used in this study. From these twelve benchmarks, the authors

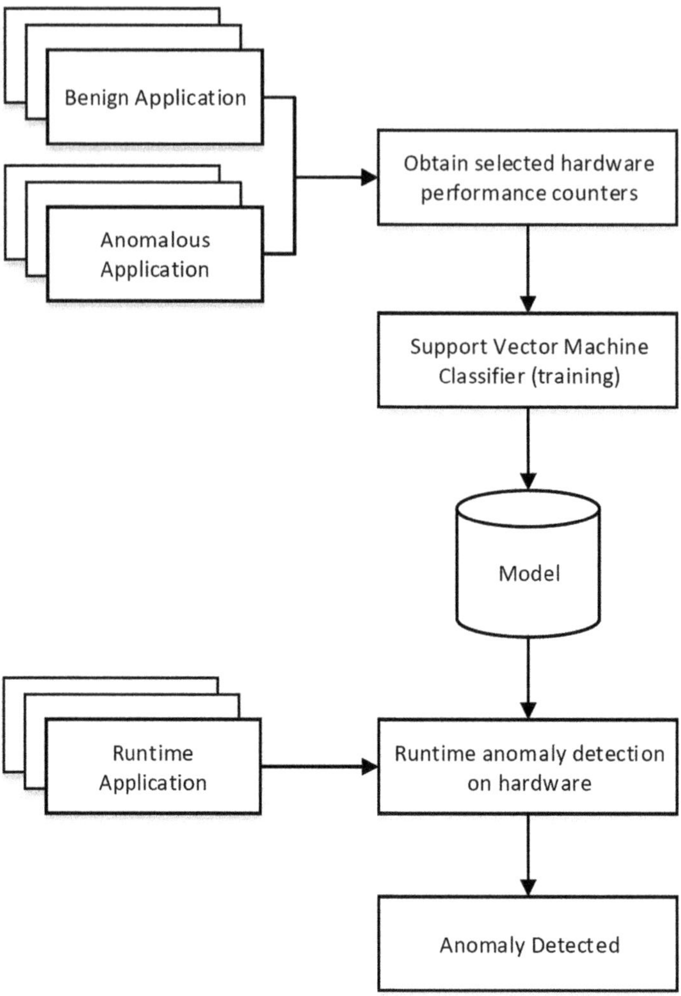


**Fig. 5.8** Methodology in case study 1, taken from [1]

generated additional thirty-one additional applications using code stitching method. The thirty-one new applications are divided into two sets: the first set contained twenty applications of two different benchmarks that are randomly mixed and the second set contained eleven benchmarks that have been patched with *reset* function. All thirty-one applications will be labelled as anomalous applications during SVM classifier training and testing phase. Standard input data provided by [12] are used for evaluation.

- **Hardware Performance Counters**:
  Once the benchmarks have been identified, the next step is to determine a set of HPCs that are suitable for anomaly detection. Due to the sensitivity of

the counters, it is highly crucial to select HPCs that can accurately model the application. The authors had used the Perf tool in Linux to obtain the hardware performance counters for all twelve benchmark applications, essentially by simply reading the counter values from the appropriate registers. As the authors are concerned of creating significant overhead in the system by reading the counter values frequently, they have opted to rely on the overall hardware performance counter values, obtained at the end of executing each benchmark application. Each benchmark application is executed ten times, and a standard deviation is calculated for each hardware performance counter. The standard deviation for each counter is then averaged across all benchmark applications. The hardware performance counters selected as features are Instruction Cache Miss, Data Cache (Read) Miss, Branch Prediction Miss and Call Instructions as the authors found that the counters with the least standard deviation have the least fluctuations.

- **Support Vector Machine (SVM) Classifier**:
  Two different classes have been identified: anomaly and benign. A SVM training algorithm builds a model that assigns new samples to one class or the other. A model generated by SVM represents the training samples as points in space, mapped so that the training samples of different classes are distinctively divided by a gap. Test samples are then mapped into that same space and predicted to belong to a class based on which side of the gap they reside. The twelve benchmark applications and twenty generated anomalous applications were selected as the training set. The twenty generated anomalous applications are separated into two categories: (a) thirteen applications represent the combination of any two benchmark and (b) seven represents the patched modification of the benchmark applications. The remaining eleven generated anomalous applications that make up from both categories are used to test the model. The authors used *scikit-learn*, a machine learning tool for Python to develop the SVM classifier with Radial Basis Function (RBF) Kernel. The values for "C" and "gamma" were 1.0 and 100, respectively, and these were obtained empirically.

## 5.6.2 Implementation

The detection of malicious software in the main processor is performed on a dedicated hardware processor. This is to achieve a fine-grained, runtime detection with minimal performance overhead. In the study, the benchmark applications are executed on the LEON3 soft processor that becomes the main processor while the Cortex-A9 processor is configured for runtime anomaly detection. All benchmark applications and generated anomalous applications are compiled and executed on the LEON3 soft processor. The authors have chosen to monitor four events, namely *Instruction Cache Miss*, *Data Cache (Read) Miss*, *Branch Prediction Miss* and *Call Instructions* for anomaly detection. Each application is executed ten times, and the four selected event counts are recorded and average value is calculated. Next, the event counts for all the benchmark applications and twenty of the generated anomalous applications are used to train the model using the SVM classifier. The model is then stored in the Cortex-A9 monitoring processor. At runtime, when an application completed its execution, the monitoring processor then extracts information from the four hardware performance counters and feeds it into the model which has been generated offline using the SVM classifier, in order to determine if the application is benign or anomalous.

### *5.6.3 Results and Discussion*

All twelve benchmark applications that were benign were used for training, as well as the first twenty anomalous applications generated. The remaining eleven generated anomalous applications have been used for testing. The benign applications were also tested using the generated model to ensure that it is correctly classified as benign. The results also show that the generated SVM model is able to distinguish between benign applications and anomalous applications although the event counts from the anomalous applications were very close to the similar benign applications.

### *5.6.4 Limitations*

While the study successfully identified anomalous applications from benign applications, the detection was performed offline rather than online. In a safety-critical application, this may not be favourable as the user will encounter a failure before the detector can identify that it is anomalous. The study also did not provide information on how long the detection process takes before it identifies whether the application is benign or anomalous. This information may be a critical factor in determining how successful the detector is in detecting anomalies.

## 5.7 Case Study 2: Anomaly Detection Using Clustering-Based Method

In this second case study, an unsupervised anomaly detection technique with Hardware Performance Counters (HPCs) to detect malware attacks is used. This case study is based on the work of [10], where the author aimed to use captured HPCs to detect malware attacks directly, without using any other statistical construction.

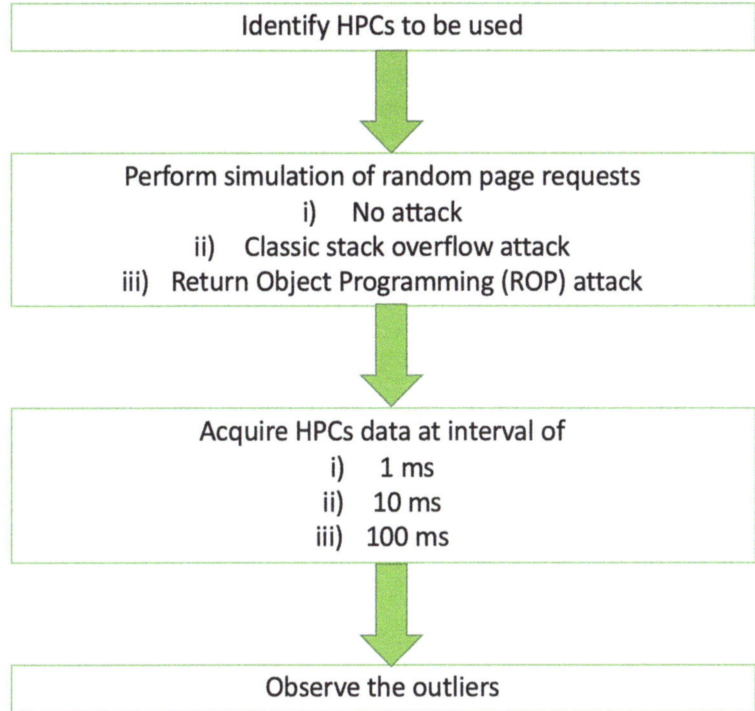


**Fig. 5.9** Methodology in case study 2

### *5.7.1 Methodology*

Figure 5.9 outlined the methodology used in the second paper, which will be further explained below:

- **Hardware Performance Counters**:
  As the number and type of events are CPU dependent, the study only focused on choosing events that are common in most CPUs. Some of the events that were considered by the author are *cpu-cycles*, *instructions*, *cache-references*, *cache-misses*, *branches*, *branch-misses*, *bus-cycles*, *ref-cycles*, *L1-dcache-loads*, *L1-dcache-stores*, *L1-icache-loads* and *L1-icache-load-misses*. The counter for the events was obtained using Linux *perf* kernel utility.
- **Data Acquisition**:
  Data are acquired by simulating a regular web browsing while executing some random page requests. In order to set up a controlled environment for the experiments, the author used a simple but real program named *nweb* from [11]. *nweb* is a tiny web server with modifications where the most important feature is that the web server has been modified to be single threaded in order to make it easier to measure the HPC. A function named *logger()* in the web software

is vulnerable to a stack overflow attack. After page requests have been made, the attack was performed. The data generated at every execution contained the timestamp of the event, the event measure delta and the name of the event. The experiments are repeated at every 1 ms, 10 ms and 100 ms.

- **Clustering Approach**:
  The author decided to use unsupervised anomaly detection technique to detect outliers. Clustering is to group similar set of objects, also known as clusters, in a dataset. There are several clustering methods such as connectivity based, centroid based, distribution based and density based, and the author has decided to apply density-based clustering method to detect malware attacks. One form of density-based clustering approach is using Local Outlier Factor (LOF). LOF tries to find anomalies by measuring the local deviation of a given data object with respect to its neighbours. LOF is based on the concept of a local density, where the locality is given by k nearest neighbours. Objects that have local densities as their neighbours will be grouped together, while objects that have lower density than their neighbours will be identified as outliers.

### 5.7.2 *Implementation*

The experiment was performed using *nweb*, where the author executed some random page requests for a while as to simulate regular web browsing and finally, after the page requests, the attack was performed. Two types of common attacks are used: (a) stack overflow and (b) Return-Oriented Programming (ROP). Each experiment is executed in three ways: (a) clean exit (no attack), (b) stack overflow attack and (c) ROP attack. The HPC counter values are collected using *perf* kernel utility tool at every 1-ms, 10-ms and 100-ms interval.

### 5.7.3 *Results and Discussion*

From the results obtained, the author showed that it is possible to identify either stack overflow attack or ROP attack using LOF density-based clustering without the need of additional statistical tools. The study also discovered that reading HPC counter values every 100 ms improved the efficiency of the LOF algorithm and reading the counter at every 1 ms and 10 ms did not provide any additional benefit. The algorithm also produced better results when using 5 neighbour points ($k = 5$) in the LOF algorithm. From the data analysis and comparison, six candidate counters were found: *iTLB-loadmisses*, *dTLB-loads*, *bus-cycles*, *LLC-store-misses*, *LLC-loads* and *LLC-load-misses*. Out of these six counters, five are related to cache operations. *iTLB* refers to the instruction cache while *dTLB* is about data cache in the Translation Lookup Buffer (TLB), and *LCC-** counters refer to the last level of the cache hierarchy.

### *5.7.4 Limitations*

In this study, while the author has showed that it has successfully distinguished malware attacks from normal web browsing, it was performed on data that were collected offline, rather than online detection. There was also no measurement of accuracy to determine how well the anomaly detection was performed. This information is crucial and is often the success criteria in determining the performance of the anomaly detection technique.

## 5.8 Conclusion

This chapter presents an overview of Hardware Performance Counters (HPCs) and how HPCs can also be used to detect anomalous behaviour in the system besides performance monitoring. By utilising the built-in HPC, the overhead incurred will be lower than for software profilers [24].

## References

1. M.F.B. Abbas, S.P. Kadiyala, A. Prakash, T. Srikanthan, Y.L. Aung, Hardware performance counters based runtime anomaly detection using SVM, in *TRON Symposium (TRONSHOW)*, pp. 1–9 (2017). https://doi.org/10.23919/TRONSHOW.2017.8275073
2. ARM, Cortex-A9 Technical Reference Manual (2009)
3. M.B. Bahador, M. Abadi, A. Tajoddin, HPCMalHunter: Behavioral malware detection using hardware performance counters and singular value decomposition, in *4th International Conference on Computer and Knowledge Engineering (ICCKE)*, pp. 703–708 (2014). https://doi.org/10.1109/ICCKE.2014.6993402
4. S. Behling, R. Bell, P. Farrell, H. Holthoff, F. O'Connell, W. Weir, The POWER4 Processor Introduction and Tuning Guide. IBM (2001)
5. S.V. Browne, J. Dongarra, N. Garner, K.S. London, P.J. Mucci, A scalable cross-platform infrastructure for application performance tuning using hardware counters, in *Proceedings of the 2000 ACM/IEEE Conference on Supercomputing, SC 2000*, p. 42 (IEEE Computer Society, Dallas, TX, USA, 2000)
6. M. Chiappetta, E. Savas, C. Yilmaz, Real time detection of cache-based side-channel attacks using hardware performance counters. Appl. Soft Comput. **49**(C), 1162–1174 (2016). https://doi.org/10.1016/j.asoc.2016.09.014
7. W.E. Cohen, Tuning programs with OProfile. Wide Open Mag. **1**, 53–62 (2004)
8. S. Das, J. Werner, M. Antonakakis, M. Polychronakis, F. Monrose, SoK: The challenges, pitfalls, and perils of using hardware performance counters for security, in *2019 IEEE Symposium on Security and Privacy, SP (2019)*. https://doi.org/10.1109/SP.2019.00022
9. J. Demme, S. Sethumadhavan, Rapid identification of architectural bottlenecks via precise event counting. SIGARCH Comput. Archit. News **39**(3), 353–364 (2011). https://doi.org/10.1145/2024723.2000107
10. A. Garcia-Serrano, Anomaly detection for malware identification using hardware performance counters. ArXiv:abs/1508.07482 (2015)

11. N. Griffiths, nweb: A tiny, safe web server (2015). https://www.ibm.com/developerworks/systems/library/es-nweb/
12. Y. Hara, H. Tomiyama, S. Honda, H. Takada, K. Ishii, CHStone: A benchmark program suite for practical c-based high-level synthesis, in *ISCAS* (IEEE, 2008), pp. 1192–1195
13. D. Heller, Rabbit: a performance counters library for Intel/AMD processors and Linux (2000)
14. Intel, Intel ®64 and IA32 Architectures Performance Monitoring Events. Intel (2017)
15. V. Jyothi, X. Wang, S.K. Addepalli, R. Karri, BRAIN: Behavior based adaptive intrusion detection in networks: Using hardware performance counters to detect DDoS attacks, in *29th International Conference on VLSI Design and 15th International Conference on Embedded Systems (VLSID)*, pp. 587–588 (2016). https://doi.org/10.1109/VLSID.2016.115
16. E.W.L. Leng, M. Zwolinski, B. Halak, Hardware performance counters for system reliability monitoring, in *IEEE 2nd International Verification and Security Workshop (IVSW)*, pp. 76–81 (2017). https://doi.org/10.1109/IVSW.2017.8031548
17. C. Malone, M. Zahran, R. Karri, Are hardware performance counters a cost effective way for integrity checking of programs, in *6th ACM workshop on Scalable Trusted Computing* (ACM, 2011), pp. 71–76
18. W. Mathur, J. Cook, Toward accurate performance evaluation using hardware counters, in *ITEA Modeling and Simulation Workshop* (2003)
19. K. Ott, R. Mahapatra, Hardware performance counters for embedded software anomaly detection, in *2018 IEEE 16th International Conference on Dependable, Autonomic and Secure Computing, 16th International Conference on Pervasive Intelligence and Computing, 4th International Conference on Big Data Intelligence and Computing and Cyber Science and Technology Congress*, pp. 528–535 (2018)
20. Perfmon2, improving performance monitoring on Linux. https://perfmon2.sourceforge.net/
21. Processor counter monitor, https://github.com/opcm/pcm
22. S. Rasoolzadeh, M. Saedpanah, M.R. Hashemi, Estimating application workload using hardware performance counters in real-time video encoding, in *7th International Symposium on Telecommunications (IST)*, pp. 307–311 (2014). https://doi.org/10.1109/ISTEL.2014.7000719
23. E. Rohou, Tiptop: Hardware performance counters for the masses, in *2012 41st International Conference on Parallel Processing Workshops*, pp. 404–413 (2012)
24. H. Sayadi, H.M. Makrani, S.M.P. Dinakarrao, T. Mohsenin, A. Sasan, S. Rafatirad, H. Homayoun, 2SMaRT: A two-stage machine learning-based approach for run-time specialized hardware-assisted malware detection, in *Design, Automation Test in Europe Conference Exhibition (DATE)*, pp. 728–733 (2019). https://doi.org/10.23919/DATE.2019.8715080
25. R. Schöne, D. Hackenberg, On-line analysis of hardware performance events for workload characterization and processor frequency scaling decisions, in *Proceedings of the 2nd ACM/SPEC International Conference on Performance Engineering* (Association for Computing Machinery, New York, NY, USA, 2011), pp. 481–486. https://doi.org/10.1145/1958746.1958819
26. B. Sprunt, The basics of performance-monitoring hardware. IEEE Micro **22**(4), 64–71 (2002)
27. SUN Microsystems, UltraSparc™ User Manual. SUN Microsystems (1997)
28. Technologies for measuring software performance (2003)
29. L. Uhsadel, A. Georges, I. Verbauwhede, Exploiting hardware performance counters, in *5th Workshop on Fault Diagnosis and Tolerance in Cryptography (FDTC)*, vol. 00, pp. 59–67 (2008). https://doi.org/10.1109/FDTC.2008.19
30. S. Wang, W. Zhang, T. Wang, C. Ye, T. Huang, VMon: Monitoring and quantifying virtual machine interference via hardware performance counter, in *IEEE 39th Annual Computer Software and Applications Conference*, vol. 2, pp. 399–408 (2015). https://doi.org/10.1109/COMPSAC.2015.14
31. X. Wang, R. Karri, Reusing hardware performance counters to detect and identify kernel control-flow modifying rootkits. IEEE Trans. Comput. Aided Des. Integr. Circuits Syst. **35**(3), 485–498 (2016). https://doi.org/10.1109/TCAD.2015.2474374

32. X. Wang, C. Konstantinou, M. Maniatakos, R. Karri, S. Lee, P. Robison, P. Stergiou, S. Kim, Malicious firmware detection with hardware performance counters. IEEE Trans. Multi-Scale Comput. Syst. **2**(3), 160–173 (2016). https://doi.org/10.1109/TMSCS.2016.2569467
33. T. Willhalm, R. Dementiev, Intel®performance counter monitor - a better way to measure CPU utilization (2012). www.intel.com/software/pcm
34. L.L. Woo, M. Zwolinski, B. Halak, Early detection of system-level anomalous behaviour using hardware performance counters, in *Design, Automation Test in Europe Conference Exhibition (DATE)*, pp. 485–490 (2018). https://doi.org/10.23919/DATE.2018.8342057

# Chapter 6
# Anomaly Detection in an Embedded System

**Lai Leng Woo, Mark Zwolinski, and Basel Halak**

## 6.1 Introduction

RazakSAT, as shown in Fig. 6.1, is an earth observation satellite that was launched on July 14, 2009. It was the first satellite in the world placed into a Near-Equatorial Low-Earth Orbit (NEqO), providing many imaging opportunities for countries around equatorial region, such as Malaysia. It was targeted to have an operational lifespan of 3 years, however, it ceased operation on August 30, 2010, just a year and 16 days from the launch date. The NEqO orbit exposes the satellite to the South Atlantic Anomaly (SAA) phenomenon on every orbit it takes around the earth. SAA is a region of reduced magnetic intensity where the inner radiation belt makes its closest approach to the Earth's surface. Satellites in low-Earth orbit pass through the SAA periodically, exposing them to several minutes of strong radiation each time, creating problems for scientific instruments, human safety and single event upsets (SEU) [4]. The failure of RazakSAT resulted in a loss of RM10.89 million in 2009, of which RM7.7 million went towards insurance premiums for the faulty satellite [40].

L. L. Woo (✉)
Malaysian Space Agency (MYSA), National Space Centre, Banting, Malaysia
e-mail: elena@mysa.gov.my

M. Zwolinski · B. Halak
School of Electronics and Computer Science, Faculty of Engineering and Physical Sciences, University of Southampton, Southampton, UK
e-mail: mz@ecs.soton.ac.uk; basel.halak@soton.ac.uk



B. Halak (ed.), *Hardware Supply Chain Security*,
https://doi.org/10.1007/978-3-030-62707-2_6

**Fig. 6.1** RazakSAT satellite, source from https://www.angkasa.gov.my

The Therac-25 was a computer-controlled radiation therapy machine produced by Atomic Energy of Canada Limited (AECL) in 1982. It suffered a concurrent programming error which saw the system giving its patients radiation doses that were hundreds of times greater than normal, thus resulting in death or serious injury [28].

On June 4, 1996, the maiden flight of Ariane 5 launcher, known as Flight 501, veered off its flight path, broke up and exploded about 40 s after the initiation of the flight sequence. The end result was that the entire mission was a failure and the cost which includes the destroyed spacecrafts was approximately $370 Million. The report issued by the Inquiry Board in charge of inspecting the Ariane 5 Flight 501 failure concluded that the failure of the active and back-up Inertial Reference System caused the two solid boosters to steer or swivel into extreme positions, and slightly later, the Vulcain engine swivelled, causing the launcher to veer abruptly [29].

RazakSAT, Therac-25 and Ariane 5 Flight 501 are examples of failure in a critical embedded system. A critical system can be divided into three categories:

1. Mission-Critical Systems: A system whose failure may result in the failure of some goal-directed activity. Some examples of mission-critical systems are an on-board computer or a navigational system in a spacecraft.
2. Business-Critical Systems: A system whose failure may result in very high costs for the business using that system. Examples of business-critical systems are the customer accounting system in a bank or the online shopping cart.
3. Safety-Critical Systems: A system whose failure may result in loss of life, injuries or significant damage to property or the environment.

The improvement on transistors size and integrated circuit performance, known as technology scaling, has allowed the growth of these computing systems across various missions [12]. Technology scaling has set the pace for semiconductor industries over the last decade whereby, with every technology generation, it had resulted in lower cost, lower power consumption, higher performance and higher transistor density per die but it also came with a cost: cheaper and better performance transistors are becoming less and less reliable [12, 43].

The failure observed in incidents, like RazakSAT, Therac-25, Ariane 5 Flight 501 or in other similar examples, had a devastating impact not only to society in general but also to the economy and environment of a country. The presence of anomalies in the systems had gone undetected, and users were only aware that something had gone wrong when a failure occurred. Research in fault prevention looks into ways of strengthening the circuit, architecture or even system from these issues. Fault prevention techniques are applied during the design and manufacturing phases with the focus on designing a better circuit, a better architecture or a better system to prevent fault [5] with techniques such as radiation hardening, shielding or modifying existing circuits or architectures.

Another way to attain dependability for systems that are already in operation is through fault tolerance. Fault tolerance research is about preventing system failure in the presence of errors, and it is usually achieved through the implementation of redundancy, error detection and recovery [5, 20]. However, current research on fault tolerance only look at detecting errors through the failure that occurs, but to date, there had been no research on predicting potential failures in real-time by detecting anomalous behaviour in the system before the user encounters the failure.

## 6.2 Chapter Overview

This chapter is organised as follows. Section 6.3 presents three type of anomaly detection techniques available, while Sect. 6.4 presents a case study of using a single Hardware Performance Counter (HPC) to predict potential failure in an embedded system by monitoring and detecting anomalous behaviour in the system, while Sect. 6.5 concludes the whole chapter.

## 6.3 Techniques for Anomaly Detection

Anomaly detection is one of many approaches used for online error detection in an embedded system. It is about detecting "likely errors" by detecting anomalous behaviours. Some examples of detectable anomalies in a microcontroller are unusual data values, branch mispredictions, exceptions, page faults, crashes, etc.

Due to the nature of data and the type of anomalies that occur in different application domains, applying an anomaly detection technique developed for one domain into another domain is not a straightforward task [7]. Availability of labels

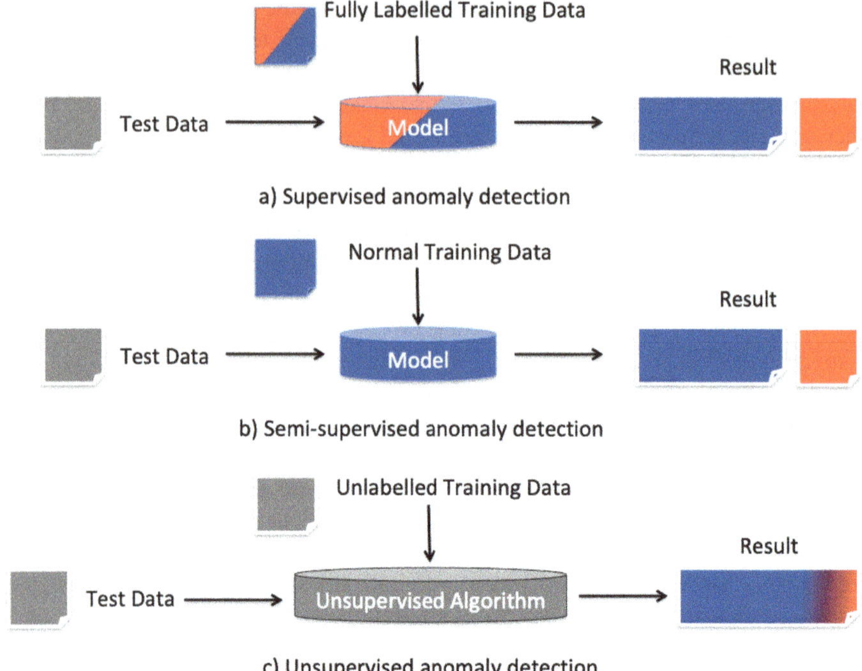


**Fig. 6.2** Different anomaly detection modes depending on the availability of labels in the dataset. (**a**) Supervised anomaly detection uses a dataset that contains both normal and anomalies for training. (**b**) Semi-supervised anomaly detection uses a "normal behaviour" training dataset. (**c**) Unsupervised anomaly detection algorithms use unlabelled dataset for training

in the dataset also plays an important role in deciding the type of techniques to be used. There are numerous techniques proposed for anomaly detection such as Replicator Neural Networks (also known as Autoencoders), One-Class Support Vector Machines, Bayesian Networks, Hidden Markov Models (HMMs), K-based Nearest Neighbours, Fuzzy Logic-based techniques and many more. Choosing an appropriate technique for anomaly detection depends very much on the available dataset as illustrated in Fig. 6.2. In general, anomaly detection techniques can be divided into three broad categories as below [18]:

- ***Supervised anomaly detection***:
  Techniques that fall under the category of supervised anomaly detection require the available samples in each dataset to be labelled as either "normal" or "abnormal". A common approach involves training a classifier and building a predictive model for both normal and abnormal class. New, unobserved data is then compared to the model to determine which class it belongs to. Some widely used techniques for supervised anomaly detection are Support Vector Machine (SVM) [1], decision trees, logistic regression, multi-layer perceptron networks

[25] and linear regression. However, there are some issues in using this approach as discussed in [7, 25]. Firstly, the available data are imbalanced as anomalous data is harder to obtain and less frequently compared to normal data. Imbalanced data causes overfitting and the trained model lacks generalisation. Another issue is the challenges faced in obtaining labelled data, especially for anomalous data. Labelling the data requires a human expert, and it is a costly exercise, not to mention time-consuming, to obtain a huge amount of all possible samples of anomalous data.

- ***Unsupervised anomaly detection***:
  Techniques for unsupervised anomaly detection do not require any training data. The data samples in the training set could contain both normal and anomalous data. However, it makes the assumption that there are more normal data instances compared to anomalous data in the training set. If the assumption is incorrect, then these techniques suffer from high false alarm rate [7]. Clustering-based techniques [13], Hierarchical Temporal Memory (HTM) networks [2], Principal Component Analysis (PCA) [36], one-class SVM and Self-Organising Map (SOM) are some techniques used for unsupervised anomaly detection.
- ***Semi-supervised anomaly detection***:
  For techniques that fall under this category, it involves building a model representing normal behaviour from a given *normal* training dataset [37]. As it does not require labels from anomalous class, this approach is more widely used compared to supervised techniques. Unknown samples are classified as outliers when their behaviour is far from that of the known normal samples. Common techniques for semi-supervised anomaly detection include statistical-based techniques [17, 21, 26, 41], one-class classifiers [23], cluster-based techniques [42], probability density function and others.

Supervised techniques require both normal and anomalous data in building a model. Imbalanced data will give a low accuracy to the model. As the amount of anomalous data available from the dataset is less than 10% while the majority of the data consist of points that depict a normal behaviour, supervised techniques for anomaly detection are not suitable and thus, will not be considered in this thesis. Unsupervised techniques can be effective only if the assumption of having more "normal" data holds true. Else, a high false detection will occur. Semi-supervised techniques which use a normal training dataset to build a model is a more balanced approach compared to supervised and unsupervised techniques. According to [14], using this approach, it is possible to achieve high detection rate with good accuracy. Although there exist numerous techniques for semi-supervised anomaly detection, not all techniques are suitable for detecting anomalies online and in real-time [2]. Most of the techniques used in real-time streaming time-series data are statistical techniques that are computationally lightweight, as one of the main requirements is the ability of the algorithm to learn continuously without storing the whole stream of data [38].

## 6.4 A Case Study of Using HPCs for Early Detection and Prediction of Failure

A dedicated hardware-based detector can be expensive and intrusive, while a pure software-based detector, though unobtrusive, may be too slow to react. Existing online error detection techniques look at detecting errors through the failures they encounter, and very often, users are only aware of the anomalous behaviour after a failure has occurred. Hardware Performance Counters (HPCs) have been shown to be used to profile a behaviour of a system, and any deviation from the normal behaviour profile indicates anomalous behaviour. This section presents a case study of using HPCs for early detection and prediction of failure in an embedded system. Anomaly detection is based on the concept of modelling what is normal in order to discover anything and everything that is different. This novel algorithm is able to predict potential failure in real-time by detecting anomalous behaviour in a processor using a single HPC.

### *6.4.1 Identification of Anomalous Behaviour*

As shown in [27], the behaviour of a system can be modelled using a single Hardware Performance Counter (HPC). As the number of available counters in a processor is limited, this imposes a limit on the number of Performance-Monitoring Events (PMEs) that can be monitored concurrently in real-time. As shown by researchers in [6, 10], using a single counter to monitor a single PME is sufficient to describe the behaviour of a program. Due to the limitation of available counters for monitoring, selection of the PME is important to ensure it is applicable across benchmarks that have different instruction distributions and that run on different processors. Architectural PMEs are the common events that can be monitored across different processors and architectures. Two different PMEs, namely the number of instructions retired (IR) and the number of cache misses (CM), are chosen as these two PMEs are common events that are available in most processors.

Several benchmarks from MiBench [19] were used, including (a) *Bitcount* and *QSort* from the Automotive and Industrial Control suite, (b) *Dijkstra* from the Network suite, (c) *StringSearch* from the Office suite and (d) *FFT* from the Telecommunication suite. All these benchmarks have different instruction distributions. Fault injection on these benchmarks, in the form of single bit-flips, was performed using GemFI [33], a fault injection tool based on the instruction set simulator Gem5. GemFI is, in principle, able to support any processor model and Instruction-Set Architecture (ISA) available in Gem5, although at the time of the experiment, only the Alpha and Intel x86 ISAs were available. In this study, Intel x86 ISA is used. GemFI can be used either in *System-call Emulation* (SE) or *Full-System* (FS), where in SE mode, users can emulate most common system calls, thus avoiding the need to model devices or even an OS. In FS mode, GemFI models the complete system including the OS and devices, executing both user-level and kernel-level instructions. GemFI injects faults into a processor core while simulating both user-level and kernel-level instructions and can model a complete system including the CPU, memory and the peripheral devices. Single bit-flip faults can be injected at various locations such as: (a) the *Fetch* instruction, (b) the selection of read/write registers during *Decoding* stage, (c) the result of an instruction in an *Execution* stage or (d) memory transactions during *Load and Store*.

#### 6.4.1.1 Experimental Setup

To extract the HPC data that will be used to monitor system reliability, there are several steps involved:

1. Set up the benchmarks required for testing.
   Each benchmark is compiled dynamically in two different versions—one in the original version and another with GemFI intrinsic functions added. Both versions are compiled for Intel x86 ISA. For *Bitcount*, the input to the benchmark is an array of integers, while for *QSort*, the input to the benchmark contains a list of words. The input for *Dijkstra* benchmark is a large graph in the form of an adjacency matrix, whereas for *FFT*, the input to the benchmark is an array of 32,768 floating point data. Lastly, the input for the *StringSearch* benchmark is a list of phrases.
2. Perform the simulation.
   Simulations of the benchmarks were performed in GemFI under FS mode. FS mode simulates the execution of the benchmark in an OS-based simulation environment. A script file is created to assist in the execution of the benchmarks. After fault injection has been initialised and enabled, a set of faults is created using the fault generator in GemFI. A fault configuration file describing the faults to be injected is provided for GemFI. This file is parsed at start-up and each fault is injected into one of the four internal queues, each of which corresponds to a pipeline stage. The simulation continues as normal until it is time for the fault to be injected. Once the fault has been injected (i.e. a bit has been flipped), the simulation continues. If an injected fault is activated or manifested as an error, it leads to the system experiencing some form of failure, else, the experiment terminates successfully. Each experiment is executed in six conditions:

   (a) *Initial Run* refers to running the binary executables without any GemFI API functions added to it. The *Initial Run* condition was executed to obtain the original behaviour of the benchmarks without any GemFI API functions added to it.
   (b) *With Fault Activated* refers to running the binary executables that have been added with GemFI API functions, but fault is not being injected. This condition provides the baseline behaviour for all benchmarks
   (c) *Fault injection in the Fetch Instruction* refers to running the binary executables with GemFI API functions added, and a fault is injected in the Fetched instruction.

(d) *Fault injection in the read or write register during Decoding stage* refers to running the binary executables with GemFI API functions added, and a fault is injected in either a read or a write register during Decoding stage.
(e) *Fault injection in the result of an instruction during Execution stage* refers to running the binary executables with GemFI API functions added, and a fault is injected in a register that contains the result of an instruction at the Execution stage.
(f) *Fault injection during memory transactions in Load/Store stage* refers to running the binary executables with GemFI API functions added, and a fault is injected in the memory instruction that is either loading a value from a register or storing a value into a register.

For each experiment, the fault model applied is a single bit-flip fault model where a single bit-flip fault is injected randomly at a stage in the pipeline. The runtime for each experiment ranges from a minimum of 5 min–2 h depending on the benchmark, the clock speed and the sampling interval. Lower clock speeds and smaller sampling intervals result in longer runtime and generation of a huge amount of data. For example, for the FFT benchmark running at clock speed 250 MHz and sampling at 5000 ns, the total runtime required was 2 h and the total data generated was 8 GB. Compare with the same benchmark but running at a clock speed of 2 GHz and sampling at 100,000 ns, the total runtime required was just around 30 min and the total data generated was 100 MB. For each experiment conducted, the HPCs are traced using the method outlined below.

3. Trace and record the required HPC values.

   The counter value is logged at certain intervals. Using this method, the execution profile for each benchmark is created, and the execution profile is able to detect the instance an error has occurred which causes a failure to the system. Sampling interval plays an important factor in determining the accuracy of time sampling methods. It is important to ensure that the execution profiles created contain sufficient amount of data that can be used to identify the anomalous behaviour in the system. Several sampling intervals were chosen to determine which interval duration is most suitable for this work. The sampling intervals listed below were chosen from the order of magnitude 2 to the order of magnitude 5, increasing one order of magnitude each time.

   - 100,000 ns;
   - 50,000 ns;
   - 10,000 ns;
   - 5000 ns;
   - 1000 ns; and
   - 800 ns.

4. Obtain, compare and analyse the results.

   The counter values obtained from *Initial Run* condition and *With Fault Activated* condition are first plotted and compared. Besides trying to establish the base line behaviour for each benchmark, the comparison is also done to ensure that the insertion of GemFI API will not alter the behaviour of the benchmarks. The base line behaviour is used to study and compare between two different PMEs, the various sampling interval as well as using different input data. Next, the counter values obtained from the remaining four conditions, (a) *Fault injection in the Fetch Instruction*, (b) *Fault injection in the read or write register during Decoding stage*, (c) *Fault injection in the result of an instruction during Execution stage* and (d) *Fault injection during memory transactions in Load/Store stage* are obtained. The counter values from the four conditions are compared against counter value obtained from *With Fault Activated* condition that forms the baseline of the behaviour for each benchmark. The outcome from each experiment are categorised either as (a) *Crash*, (b) *Hang*, *Fail Silence Violation* or *Not Manifested*.

#### 6.4.1.2 Comparison Between Two PMEs

Figure 6.3 compares the execution profiles obtained from Dijkstra benchmark running at two clock speeds (a) 250 MHz, and (b) 2 GHz. The profiles were generated

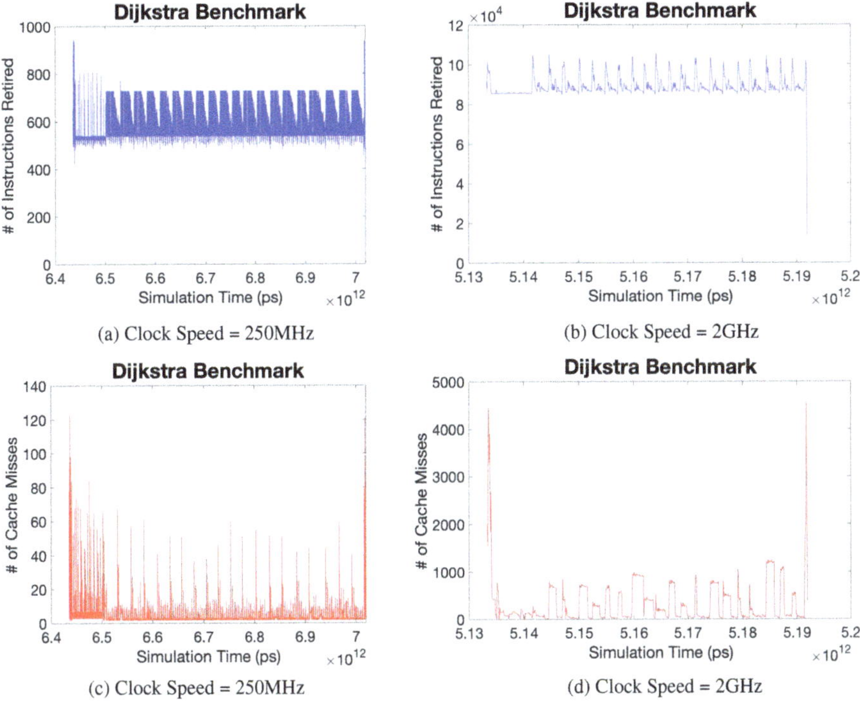


**Fig. 6.3** Execution profiles using Number of Instructions Retired and Number of Cache Misses for Dijkstra benchmark running at 250 MHz and 2 GHz clock speed. (**a**) Clock speed = 250 MHz. (**b**) Clock speed = 2 GHz. (**c**) Clock speed = 250 MHz. (**d**) Clock speed = 2 GHz

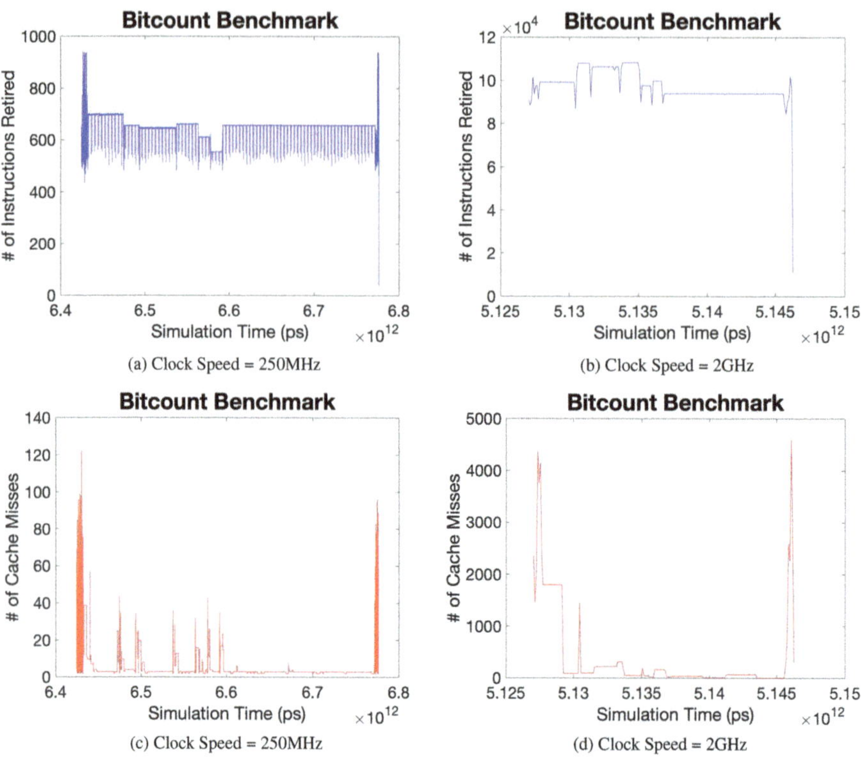


**Fig. 6.4** Execution profiles using Number of Instructions Retired and Number of Cache Misses for Bitcount benchmark running at 250 MHz and 2 GHz clock speed. (**a**) Clock speed = 250 MHz. (**b**) Clock speed = 2 GHz. (**c**) Clock speed = 250 MHz. (**d**) Clock speed = 2 GHz

from *With Fault Activated* condition (i.e. fault was not injected), which forms as the baseline behaviour for this benchmark. Figure 6.3a,b show the behaviour of Dijkstra benchmark monitored using the IR PME against the simulation time in picoseconds, while Fig. 6.3c,d show another set of results where the behaviour of the benchmarks were monitored using the CM PME plotted against simulation time in picoseconds. As can be seen in Fig. 6.3, the profile for each benchmark is similar even though it is running at a different clock speed. From the results, it shows that a program exhibits the same behaviour regardless of any clock speed it runs on. This suggests that the PME monitored using HPC can be used to create the execution profile of an application, and thus, identifying the normal behaviour of the system.

Figure 6.4 shows the execution profiles obtained from Bitcount benchmark where Fig. 6.4a,b show the behaviour of Bitcount benchmark monitored using IR PME (axis Y) against the simulation time in picoseconds (axis X), while Fig. 6.4c,d show another set of results where the behaviour of the benchmarks were monitored using CM PME plotted on axis Y. It is clear that the execution profiles for Dijkstra benchmark shown in Fig. 6.3 completely differs from the execution profiles for

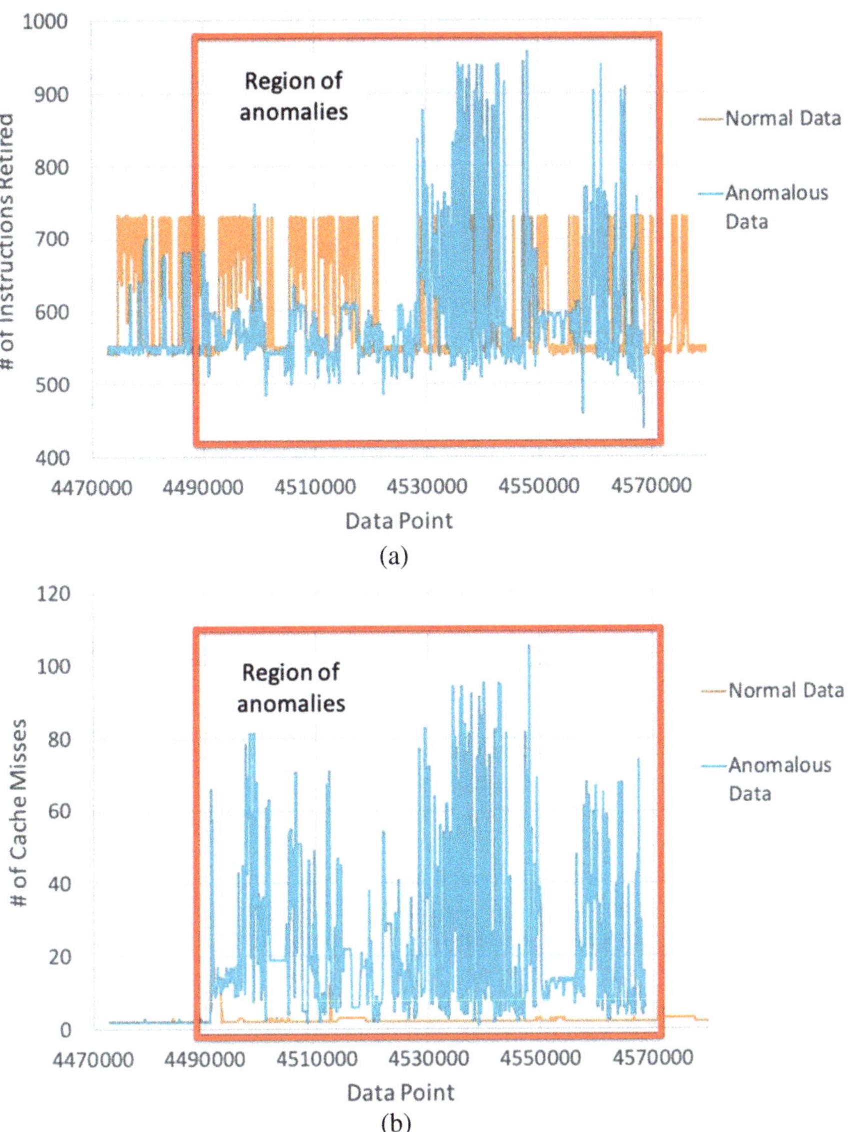


**Fig. 6.5** Comparison between Number of Instructions Retired (IR) and Number of Cache Misses (CM) for Dijkstra benchmark running at 250 MHz clock speed

Bitcount benchmark shown in Fig. 6.4 whether it is using instructions retired PME or cache misses PME. From the results obtained, it is clear that each application has its own signature profile, which can be monitored using HPC.

Figure 6.5 compares the values recorded using the CM PME and IR PME. The values recorded for CM are between 0 and 110, while those recorded for the IR PME are between 450 and 950. In general, the values recorded for CM have between three and seven bits, while the values recorded for IR have around seventeen bits, more than twice as many. Therefore, the computational size and speed can be greatly reduced by monitoring the behaviour of CM compared to IR. Another observation is that the CM is also easier to detect because the counter data has a bigger deviation, and thus provides better detection accuracy. Based on all this, it is considered that the CM PME is more suitable for monitoring anomalous behaviour in the system.

#### 6.4.1.3 Comparison on Various Sampling Interval

Figure 6.6 shows the results for Dijkstra benchmark running at 250 MHz clock speed with various sampling interval. The results display various execution profiles which were plotted with number of cache miss plotted against simulation time. The difference between each figure is the amount of data collected from the counter. For example, in Fig. 6.6d where experiment with the sampling interval set at 5000 ns, there were more data collected compared to the experiment with 100,000 ns sampling interval as shown Fig. 6.6a. A total of 120,000 data points are collected from the experiments running at 5000 ns, while only 6000 data points are collected from the experiments running at 100,000 ns.

The shorter the duration of the interval, the higher the amount of data is generated. Shorter intervals allow the anomalous behaviour to be detected earlier as the amount of data from the point the fault manifested as an error to the point where a failure occurs increases. Another observation is the value of the counter gets smaller as the sampling interval gets smaller. This finding is consistent with the behaviour of HPC itself, where in time-based sampling, the counter is incremented on an instruction-by-instruction basis until an interrupt is generated. Smaller sampling interval means the interrupt is generated quicker. However, it is found that sampling interval below 5000 ns is not suitable as it is difficult to distinguish between an anomalous point from a normal point. Another problem of having an interval that is too short is that the same activity can be recorded several times, thus inflating the sample size. Therefore, the most suitable sampling interval chosen is 5000 ns.

#### 6.4.1.4 Comparison on Using Different Input Data

The experiment was also conducted for QSort and Dijkstra benchmarks with different sets of input data and the execution profile for each benchmark was compared. An input data is defined as a file that contains data that serve as an input to a program. Executing the benchmarks with different sets of input data simulates the condition where applications do not always run on the same input data. There are three different sets of input data used for the QSort benchmark and four different sets of input data used for Dijkstra benchmark. For QSort benchmark, the first set of input data consists of words and integers, the second set of input data consists of a mixture of words, integers and floating points and the third set of input data contains

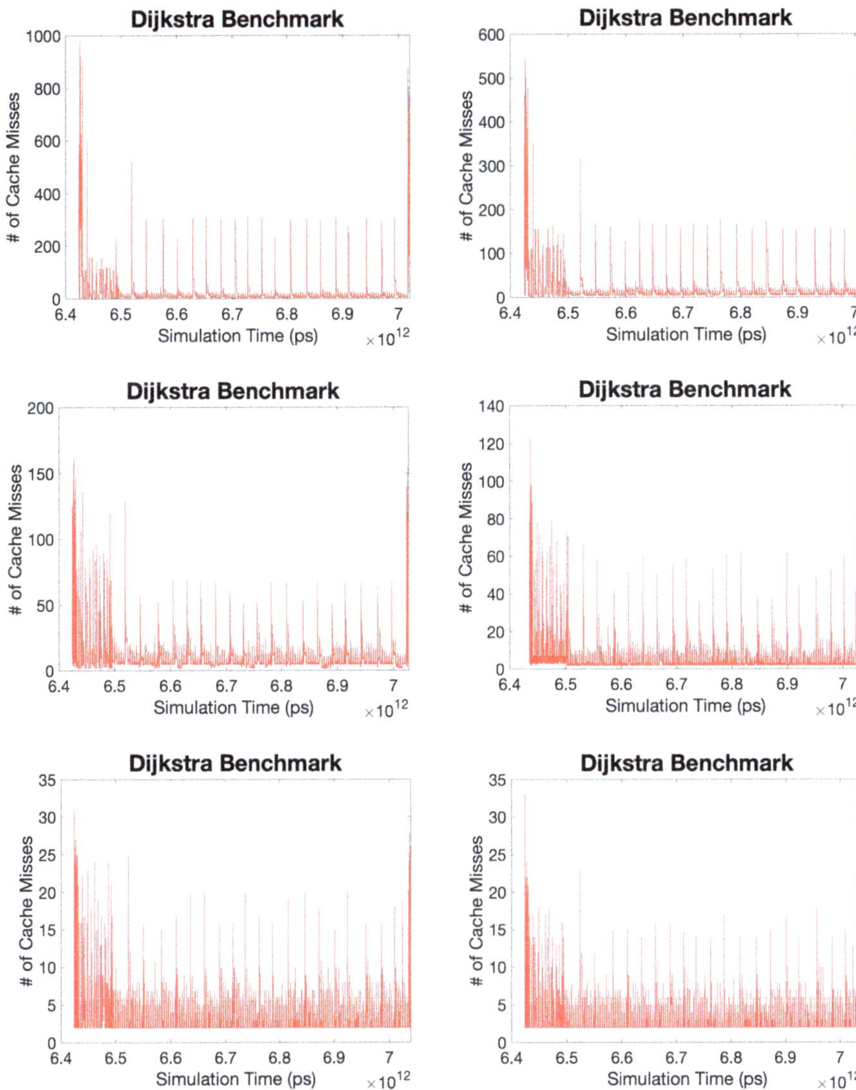


**Fig. 6.6** Execution profiles using Number of Cache Misses for Dijkstra benchmark running at different sampling interval. (**a**) 100,000 ns. (**b**) 50,000 ns. (**c**) 10,000 ns. (**d**) 5000 ns. (**e**) 1000 ns. (**f**) 800 ns

only integers and floating points. Since the Dijkstra benchmark is a benchmark that calculates the shortest path between every pair of nodes in a graph, different sets of input data means having different nodes in each input data.

Figure 6.7 shows the execution profile generated for QSort and Dijkstra benchmarks with different sets of input data and from the results shown, the execution

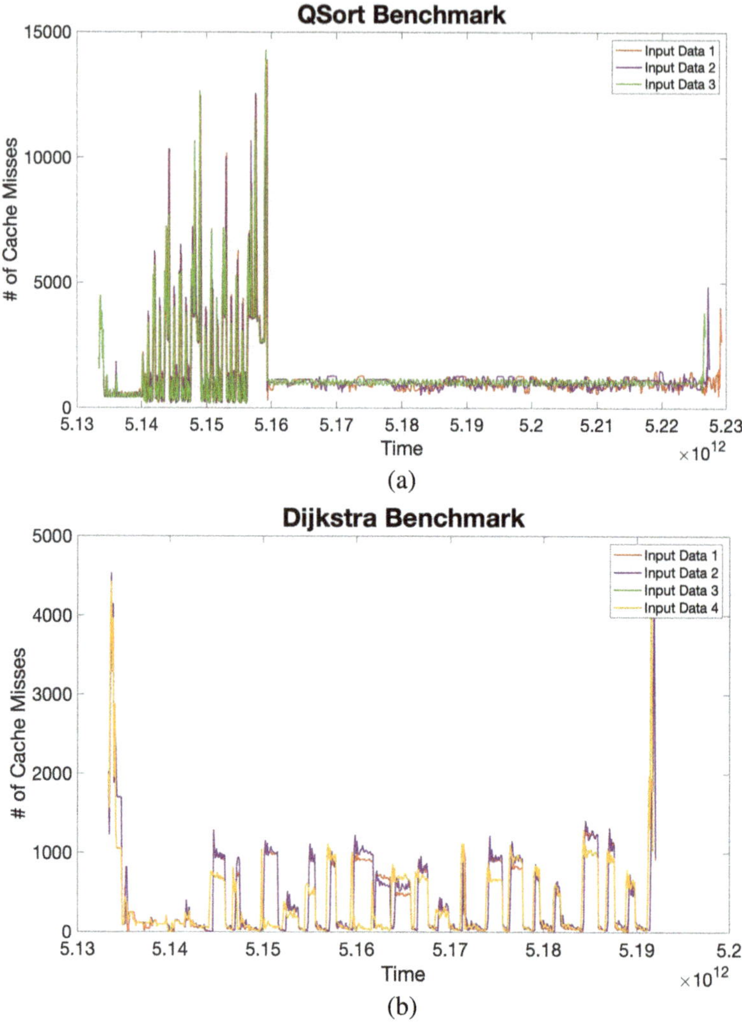


**Fig. 6.7** Execution profiles using Number of Cache Misses for (**a**) QSort and (**b**) Dijkstra benchmarks with multiple inputs.

profile for each benchmark still bears a strong similarity. This finding suggests that regardless of any input used, the execution profile remains similar, and thus, it is possible to observe anomalies that may occur based on the profiles generated by the counter.

#### 6.4.1.5 Correlation Between Errors and Failures

A failure is said to have occurred when the system is transitioned from correct service to incorrect service. A failure is caused by the presence of an error in the system where an error is the terminology used for an active or manifested fault. Fault could be due to an attack, programming issues or a defect in the hardware. To observe the correlation between errors and failures, the failures and errors are first classified and defined according to [24]. Table 6.1 described four categories of failures and Table 6.2 defines the various type of errors which may occur.

Figure 6.8 shows the distribution of failures across all five benchmarks. Only a total of 25% faults are manifested as errors and caused the system to behave anomalously. Benchmarks that contain a large percentage of arithmetic instructions such as QSort, StringSearch or FFT suffer higher percentage of "hang" failures, while memory intensive benchmarks such as Dijkstra experience "crash" failure more frequent.

**Table 6.1** Failure categories

| Failure category | Description |
|---|---|
| Crash | The system stops working |
| Hang | System resources are exhausted, resulted in a nonoperational system |
| Fail silence violation | Either the system or the application erroneously detects the violation presence of an error or allows an incorrect data/response to propagate out |
| Not manifested | The corrupted instruction is used, but it does not have a visible abnormal impact to the system. |

**Table 6.2** Error categories

| Error category | Description |
|---|---|
| Segmentation fault | Access violation, raised by hardware with memory protection, notifying an OS the software has attempted to access a restricted area of memory |
| Invalid opcode | An illegal instruction that is not defined in the instruction set is executed |
| Kernel panic | The operating system detects an error |
| NULL pointer | Unable to handle kernel NULL pointer de-reference |
| Bad paging | A page fault. The kernel tries to access some other bad page except NULL pointer |
| Assertion error | Assertion evaluates to false at runtime |
| Bad trap | Unknown exception |
| General protection fault | Exceeding segment limit, writing to a read-only code or data segment, loading a selector with a system descriptor, reading an execution-only code segment. |

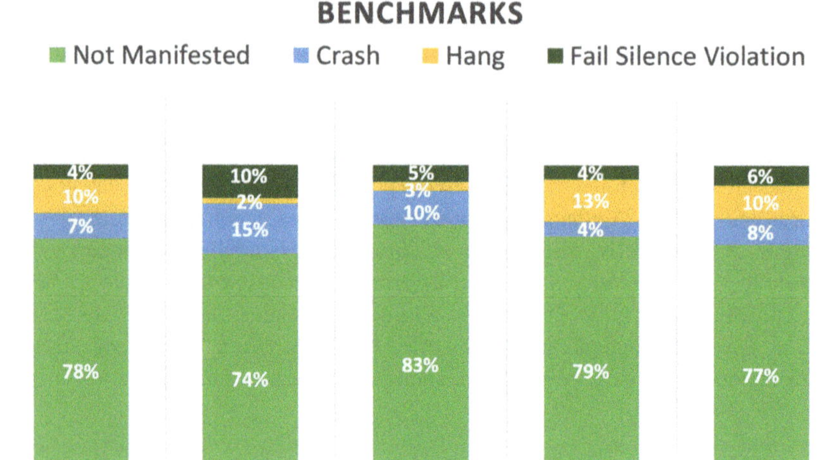


**Fig. 6.8** Failure distribution across benchmarks

#### 6.4.1.6 Normal vs Abnormal

Figure 6.9 shows how system-level anomalous behaviour can be observed as the execution profile deviates from the original profile. The Performance-Monitoring Event (PME) used to monitor the behaviour is number of cache misses. While there is a huge selection of PMEs to be used for monitoring, the number of counters available is, however, limited. Therefore, selecting the right PME is important to ensure it can be used across benchmarks that have different instruction distributions and running on different processors.

In this case study, two different PMEs were explored: the number of instructions retired (IR) and the number of cache misses (CM), to determine which PME is better for detection. The data for these two PMEs were monitored and collected separately. The benchmarks were simulated at a clock speed of 250MHz, a clock speed typically used in microcontrollers, and data was gathered every 5 $\mu$s (or 1250 clock cycles). It was found that sampling the data every 5 $\mu$s provides a sufficient amount of data for detection: more frequent sampling will generate more noise, while longer periods will suffer some loss in the data. As can be seen in Fig. 6.9, the anomalous behaviour triggered by a fault in the processor can be classified as (a) fault free (or masked error), (b) fail silence violation, (c) hang and (d) crash. These four behaviours can be observed by using just a single counter.

Figure 6.10 shows the region of collective anomalies that occurred in an anomalous dataset compared to a fault-free dataset for the Dijkstra benchmark.

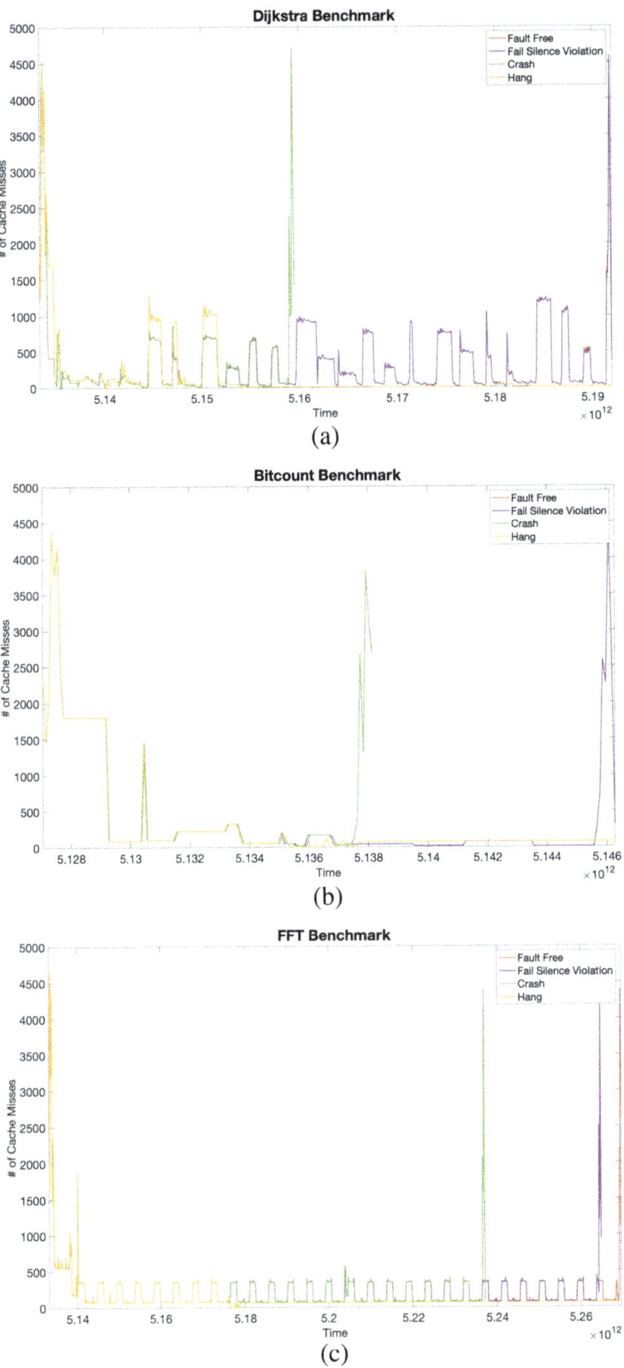


**Fig. 6.9** Anomalous behaviour detected on various benchmarks using HPC

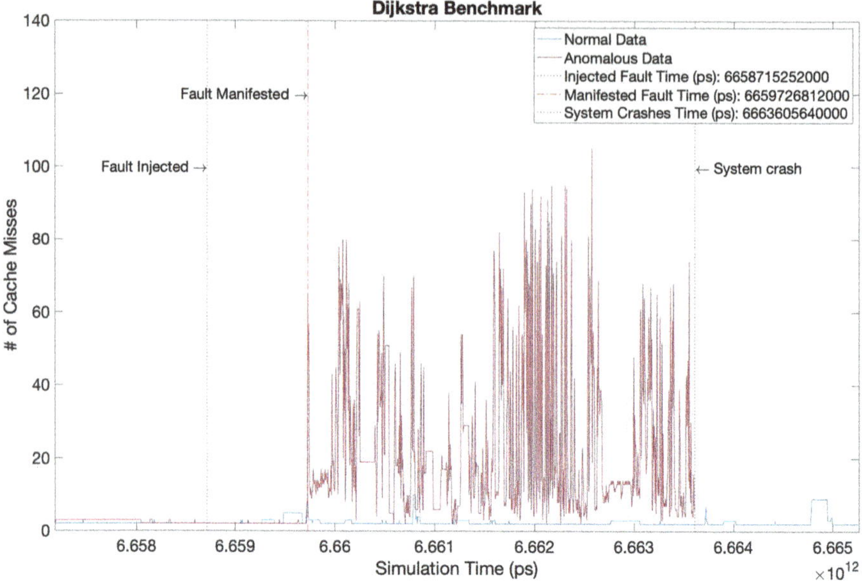


**Fig. 6.10** Occurrence of collective anomalies

When an error is present in the system, the counter values begin to deviate from the normal pattern. In the case of a *hang*, there is a huge deviation observed in the counter value at the beginning of the error before the counter value becomes constant at some point. The amount of time for the system to stay hung or unresponsive varies as it depends on when the user sends an interrupt to the system. For a *crash*, the counter will also show a huge deviation before it stops. The characteristics of the profiles for a *Masked Error* or a *Fail Silence Violation* (or *Silent Data Corruption*) are almost the same as for a fault-free system.

#### 6.4.1.7 Detection Interval

The different failures caused by manifested faults are illustrated in Fig. 6.11, which shows the temporal relationship between a fault, an error and a failure. Assuming a single bit-flip fault occurs at time $t_f$. When the fault propagates or manifests itself as an error, the counter begins to deviate, and a string of anomalies occurs from time $t_e$ until the system fails at time $t_{fail}$. The important time interval is $\delta t_d$, which is the Maximum Detection Interval between an error and a failure. For a prediction method to be useful, the detection of anomalous behaviour has to be as early as possible within the detection interval.

As shown in Fig. 6.10, when the fault is manifested as an error, the HPC data began to deviate from the normal behaviour, thus creating collective anomalies. From the time the fault is manifested as an error until the time when the failure

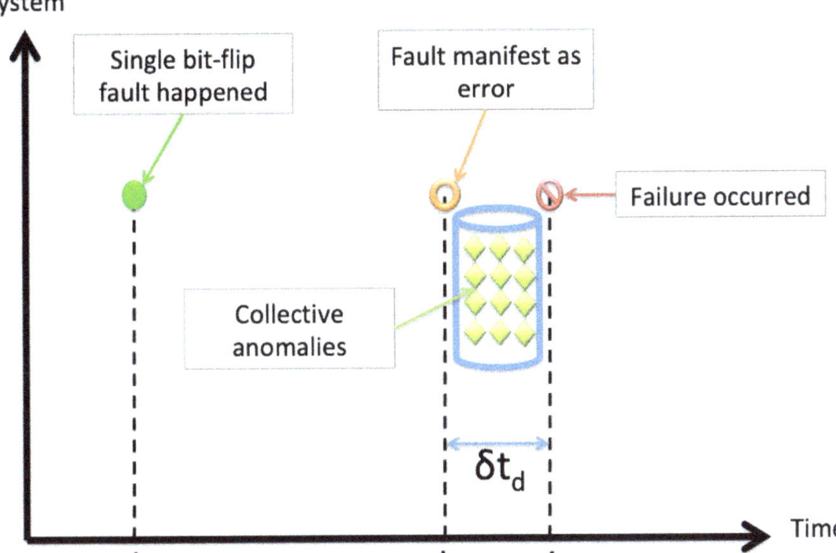


**Fig. 6.11** Temporal relationship between fault, error and failure

is observed in the system, there is a delay of approximately 1.25M clock cycles. Further experiments were conducted to determine the minimum number of clock cycles it takes for a system to crash after the fault becomes an error. It was found that the minimum amount of clock cycles it takes for a system to crash is approximately 1M, as shown in Fig. 6.12.

Therefore, $\delta t_d$ can be determined using Eq. (6.1), where the Targeted Clock Cycles is 1M and the Clock Speed and Time Unit are self-explanatory. For example, an embedded system running at 250 MHz has a maximum detection interval of 4000 μs (or 4 ms) before the system encounters a failure. Therefore, the early detection and prediction algorithm has to be able to predict a potential failure within 4000 μs.

$$\delta t_d = \frac{\text{Targeted Clock Cycles}}{\text{Clock Speed}} * \text{Time Unit} \tag{6.1}$$

## 6.4.2 Early Detection and Prediction Algorithm

An early detection and prediction algorithm is developed to detect anomalous behaviour and predict potential failure based on the characteristics observed from *hang* and *crash*. The Dijkstra benchmark is used as a case study in this section.

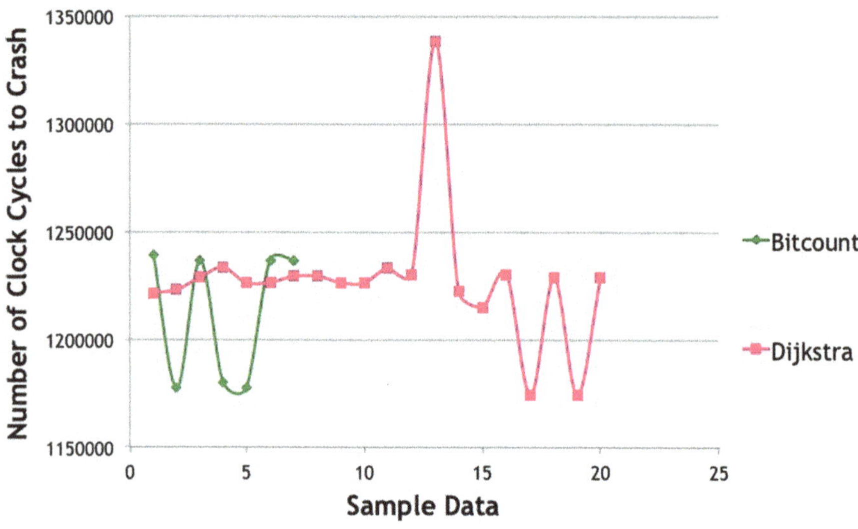


**Fig. 6.12** Number of clock cycles to crash

Based on the findings presented in Sect. 6.4.1, it is found that the CM PME is more suitable compared to IR PME for monitoring deviation in the profile. The values recorded using the CM PME are much lower, between three and seven bits. Besides that, the deviations recorded using the CM PME are also higher, making it easier to detect anomalous behaviour in the system. Hence, the CM PME is used to build the datasets, which are then used in the development of the early detection and prediction algorithm.

The sampling interval of 5000 ns is selected to generate the required datasets as it is found that this interval duration generates sufficient amount of data for the algorithm to distinguish an anomalous point from a normal point. A total of nine datasets were obtained with the input data for each dataset differ from one dataset to another. All datasets contain approximately 120,000 data points and all the datasets exhibit fault-free behaviour, which means there is no anomalous behaviour detected in the execution profile. As each dataset is different, all nine datasets will be used as training datasets. As for testing dataset, a different dataset is used. This dataset contains 118,860 data points, compared to a normal dataset that contains approximately 120,000 data points. This dataset also contains collective anomalies.

The HPC data collected from Dijkstra benchmark is a univariate time-series data, where the input data is continuously streaming at time $t$, and can be represented as $y\{t\} = y_1, y_2, y_3, \ldots, y_t$. For time-series data, a *time plot* where data are plotted over time can reveal any seasonal behaviour, trend over time or any other features of the data. To determine if the time-series is stationary, one can perform a *Unit Root Test*. The most widely used test for unit root testing is the *Augmented Dickey–Fuller* (ADF) test where the hypotheses used in this work are:

- The null hypothesis is that a unit root is present in the data; and
- The alternate hypothesis is that the time-series data is stationary.

$$\Delta Y_t = \alpha + \gamma Y_{t-1} + \delta_1 \Delta Y_{t-1} + \ldots + \delta_p \Delta Y_{t-p+1} + \epsilon_t \tag{6.2}$$

Equation 6.2 using the basic regression model that has a constant and no trend is used to test for unit root testing. The test is then carried out under the null hypothesis $\gamma = 0$ against the alternative hypothesis of $\gamma < 0$. Once the value is computed, it is compared to the relevant critical value for ADF test. If the test statistic is less than the critical value, then the null hypothesis $\gamma = 0$ is rejected and the series is a stationary series. Comparison can also be made between the calculated $DF_t$ statistic and the tabulated critical value in Table 6.3. If the calculated $DF_\tau$ is more negative than the table value, the null hypothesis is rejected. The equation for $DF_\tau$ is

$$DF_\tau = \frac{\widehat{\gamma}}{SE(\widehat{\gamma})} \tag{6.3}$$

As the number of samples in each dataset is more than 500, the critical value for $T = \infty$ is chosen. The 5% critical value from Table 6.3 is chosen as the dataset contains no trend. The ADF test is performed on all nine datasets of the Dijkstra benchmark and the value of $DF_\tau$ for each sample of data is shown in Table 6.4. From the test results shown in Table 6.4, the value of $DF_\tau$ for each dataset in

**Table 6.3** Critical values for Dickey–Fuller t-distribution, source from [16]

| Critical values for Dickey–Fuller t-distribution | | | | |
|---|---|---|---|---|
| | No trend | | With trend | |
| Sample size, T | 1% | 5% | 1% | 5% |
| T = 25 | −3.75 | −3.00 | −4.38 | −3.60 |
| T = 50 | −3.58 | −2.93 | −4.15 | −3.50 |
| T = 100 | −3.51 | −2.89 | −4.04 | −3.45 |
| T = 250 | −3.46 | −2.88 | −3.99 | −3.43 |
| T = 500 | −3.44 | −2.87 | −3.98 | −3.42 |
| T = ∞ | −3.43 | −2.86 | −3.96 | −3.41 |

**Table 6.4** ADF test results for Dijkstra benchmark

| ADF test results | | |
|---|---|---|
| Sample | Sample size | $DF_\tau$ |
| Dataset 1 | 122,934 | −91.09 |
| Dataset 2 | 117,091 | −90.60 |
| Dataset 3 | 117,091 | −90.60 |
| Dataset 4 | 122,697 | −99.56 |
| Dataset 5 | 116,855 | −96.52 |
| Dataset 6 | 119,122 | −101.99 |
| Dataset 7 | 114,191 | −99.54 |
| Datasct 8 | 120,289 | −99.21 |
| Dataset 9 | 120,471 | −100.08 |

Dijkstra benchmark is found to be more negative compared to −2.86, therefore the null hypothesis $\gamma = 0$ is rejected and the time-series is found to be stationary.

Detection of anomalous behaviour using an HPC consists of three stages, with each stage building from its predecessor:

1. Predict the next values in the time-series;
2. Measure the deviation between the predicted and observed values;
3. Determine whether the observed value deviates "too much" and can be deemed anomalous.

Algorithm 1 shows the methodology for early detection of anomalous behaviour using time-series data gathered from the HPC. We use a one-step ahead prediction method to predict the next data value $\hat{Y}_{t+1}$. The predicted value will be measured against the observed value and the observed value will be classified as anomalous if it falls outside a defined threshold. An alarm will be raised if the number of consecutive anomalous points exceeds a predefined value, else the actual observed value $Y_{t+1}$ is added to the front of the series and the next sequential data is predicted.

**Algorithm 1** Methodology for early detection

```
Predict the next Ŷ_{t+1} using One-Step Ahead Prediction
Measure the deviation between the predicted value, Ŷ_{t+1} and the observed value, Y_{t+1}
Compare the deviation with a pre-determined threshold
if (Y_{t+1} out of range) then
    Mark Y_{t+1} as anomalous
    if (Consecutive Anomalies > C) then
        Predict a potential failure in the system.
    else
        Go back to Step 1
    end if
else
    Go back to Step 1
end if
```

#### 6.4.2.1 Predicting Potential Failure

The main objective of the detector is to predict a potential failure in the system before the system fails, but at the same time, avoid being overly sensitive. Based on the case study conducted, once a fault has been manifested as an error, the hardware counter data began to deviate from its normal pattern. This leads to the crux of the algorithm—the number of consecutive anomalies required to be detected before raising an alarm for a potential failure. The optimal number of consecutive anomalies, denoted by $C$, is that needed to predict a failure in the shortest time possible. In this work, it is found that the minimum value of $C$ is 4. This means the detection and prediction algorithm has to detect at least four consecutive anomalies before raising an alarm. If $C < 4$, the detector is overly sensitive so an alarm is raised on normal points that are wrongly identified as anomalies. Based on experiments, the optimal value of $C$ is found to be 5. If $C > 5$, the detection time is increased with no benefit.

#### 6.4.2.2 One-Step Ahead Prediction

To detect manifested error as early as possible and predict possible failure, it requires time-series forecasting, i.e. making prediction of the next data instance based on a model fitted to present and past data instances. Applying a threshold rule on the HPC data will not work because the anomalies that occur in HPC data are not point anomalies, but collective anomalies. Using an overall thresholding rule on the dataset will not only result in high false alarms, but it will not be able to predict a potential failure in a timely manner. In order to detect anomalies on-chip and in real-time, a one-step ahead prediction method must have minimal computational complexity and not require any preprocessing of the data. The forecast horizon is set as 1 (hence the name one-step ahead) because it allows the deviation between the predicted value and the actual value to be observed as soon as possible. A huge deviation between the predicted and actual values indicates that an anomaly has occurred. If the forecast horizon is set several time steps away, there will be a delay in detecting the anomalous behaviour as the comparison between the predicted and observed values has to be made at the same time stamp. Petropoulos et al. [34] have shown that for regular or fast-moving data, forecasting in a short term horizon is more accurate than for a longer forecasting horizon.

Recent work by Makridakis et al. [30] shows that statistical forecasting methods, such as SES and ARMA, outperform other machine learning algorithms, such as K-Nearest Neighbour Regression (KNN), Bayesian Neural Network (BNN), Support Vector Regression (SVR) and others, while at the same time having low computational requirements. Therefore, three different methods are being investigated, namely: (a) Single Exponential Smoothing (SES); (b) Autoregressive Moving Average (ARMA) and (c) Single Layer Linear Network (LN). All nine datasets are used as training data in order to find the optimal values of the required parameters for each prediction method.

1. ***Single Exponential Smoothing (SES)*** [17, 22]
   The SES method is a type of exponential smoothing prediction method that uses historical data and assigns weights to forecast future values. The one-step ahead prediction for data at time $t = t + 1$ is a weighted average of all the observations in the series $Y_1, \ldots, Y_t$. The rate at which the weights decrease is controlled by the parameter $\alpha$, also known as the smoothing parameter. Weights are decreased exponentially for data that is further in the past. In other words, the older the data, the smaller the associated weight.
   The SES is represented in a component form following [22] as it minimises the amount of memory used to store the past data. The component form representation of SES comprises a forecast equation and a smoothing equation for each of the components included. As the time-series does not exhibit any

trend or seasonal patterns, the only component included is the level, $l_t$, given in Eq. (6.4):

$$\begin{aligned} \text{Smoothing Equation:} \quad & l_t = \alpha Y_t + (1-\alpha)\, l_{t-1} \\ \text{Forecast Equation:} \quad & \hat{Y}_{t+1} = l_t \end{aligned} \tag{6.4}$$

$\hat{Y}_{t+1}$ consists of the weighted average of the most recent observation $Y_t$ and the smoothed value of the series, $l_{t-1}$. The smoothing parameter, $\alpha$, is used to smooth or dampen older observations and has a value between 0 and 1. If $\alpha$ is small (i.e. close to 0), past observations are given more weight. Conversely, if the value of $\alpha$ is big (i.e. close to 1), more weight is given to the more recent observations. In order to begin the SES prediction process, the initial smoothed value, denoted as $l_0$, needs to be estimated. According to [31], the estimation of $l_0$ in a large dataset has little relevance. Therefore, in this work, the initial forecast value, $l_0$, is set to the initial value of the time-series, $Y_0$. Once the initial smoothed value, $l_0$, has been set, it is substituted into Eq. (6.4). The forecast value for the next period, $\hat{Y}_{t+1}$, is simply the current smoothed value.
Optimisation of $\alpha$ is done by minimising the *Mean Absolute Error* (MAE). MAE is defined as the average absolute deviation between the predicted and observed values, $e_t = \left|\hat{y}_t - y_t\right|$, Eq. (6.5). The value of $\alpha$ was varied between 0 and 1 in increments of 0.1. The MAE score is calculated for each $\alpha$ for each dataset during training. The MAE scores from all nine datasets are then averaged before selecting the values of $\alpha$ that give the lowest MAE score.

$$\text{MAE} = \frac{1}{n}\sum_{t=1}^{n} e_t \quad \text{where} \quad n = \text{sample size} \tag{6.5}$$

The first step of this experiment involves obtaining the optimal value of $\alpha$. The training data used involves all nine datasets. The $\alpha$ value is varied between 0 and 1 in increments of 0.1. The MAE is obtained for each $\alpha$ value for each dataset during training, and the MAE values from all nine datasets are then averaged with the results as shown in Table 6.5. As can be seen from Table 6.5, the lowest mean MAE achieved is 0.49 when $\alpha = 0.7$.

2. ***Autoregressive Moving Average (ARMA)*** [46]
The ARMA method (also known as the Box-Jenkins method) consists of an autoregressive (AR) part and a moving average (MA) part. The AR part, which involves coefficients $\phi_t$ with $t = 1, \ldots, p$, reflects the relationship between $\hat{Y}_{t+1}$ and the past values of the time-series. The MA part, which involves coefficients $\theta_t$ with $t = 1, \ldots, q$, reflects the relationship between $\hat{Y}_{t+1}$ and the residues. The data does not require any differencing as it is found to be stationary, based on the *Augmented Dickey-Fuller (ADF)* test. Therefore, the ARMA $(p, q)$ model is used as defined in Eq. (6.6).

$$\hat{Y}_{t+1} = c + \phi_1 Y_{t-1} + \ldots + \phi_p Y_{t-p} + \epsilon_t - \theta_1 \epsilon_{t-1} - \ldots - \theta_q \epsilon_{t-q} \tag{6.6}$$

**Table 6.5** Average mean absolute error (MAE) for different $\alpha$ values in SES

| Alpha (a) | MAE | | | | | | | | | Average |
|---|---|---|---|---|---|---|---|---|---|---|
| | Dataset 1 | Dataset 2 | Dataset 3 | Dataset 4 | Dataset 5 | Dataset 6 | Dataset 7 | Dataset 8 | Dataset 9 | |
| 0.0 | 9.3505 | 9.3135 | 9.3135 | 8.2542 | 8.2847 | 8.2410 | 8.2384 | 8.2525 | 8.2379 | 8.61 |
| 0.1 | 0.6670 | 0.6155 | 0.6155 | 0.7811 | 0.7287 | 0.7082 | 0.7214 | 0.7001 | 0.7020 | 0.69 |
| 0.2 | 0.5538 | 0.5064 | 0.5064 | 0.6667 | 0.6345 | 0.6265 | 0.6397 | 0.6134 | 0.6212 | 0.60 |
| 0.3 | 0.5002 | 0.4593 | 0.4593 | 0.6114 | 0.5936 | 0.5900 | 0.6020 | 0.5742 | 0.5829 | 0.55 |
| 0.4 | 0.4668 | 0.4326 | 0.4326 | 0.5778 | 0.5687 | 0.5673 | 0.5790 | 0.5505 | 0.5585 | 0.53 |
| 0.5 | 0.4447 | 0.4148 | 0.4148 | 0.5564 | 0.5515 | 0.5518 | 0.5639 | 0.5352 | 0.5419 | 0.51 |
| 0.6 | 0.4299 | 0.4030 | 0.4030 | 0.5427 | 0.5414 | 0.5424 | 0.5545 | 0.5264 | 0.5311 | 0.50 |
| 0.7 | 0.4203 | 0.3962 | 0.3962 | **0.5356** | **0.5364** | **0.5386** | **0.5499** | **0.5227** | **0.5257** | **0.49** |
| 0.8 | 0.4153 | **0.3940** | **0.3940** | 0.5364 | 0.5398 | 0.5418 | 0.5523 | 0.5258 | 0.5275 | 0.49 |
| 0.9 | **0.4149** | 0.3962 | 0.3962 | 0.5425 | 0.5484 | 0.5499 | 0.5597 | 0.5333 | 0.5345 | 0.50 |
| 1.0 | 0.4196 | 0.4027 | 0.4027 | 0.5547 | 0.5627 | 0.5637 | 0.5731 | 0.5466 | 0.5227 | 0.51 |

where $\hat{Y}_{t+1}$ is the variable to be predicted using previous samples of the time-series, $\epsilon_t$ denotes white noise and $c$ is a constant offset.
The model's parameters, including the coefficients denoted by $p$ and $q$, are estimated based on [31]. Based on the initial experiments conducted, it is found that the detection accuracy reduces, while the detection time increases significantly for models having coefficients greater than 4. Hence, the coefficients for $p$ and $q$ are limited to 4. By varying the coefficients $(p, q)$ between 0 and 4 in increments of 1, a total of twenty-five ARMA models with different orders of model parameters were built, with the exception of (0, 0). An ARMA (0, 0) model is used on a time-series that contains basically a constant and white noise. Since the time-series obtained from the Dijkstra benchmark does not consists of a constant and white noise, ARMA (0, 0) is not considered. To assist in identifying a suitable ARMA model, the *Akaike Information Criterion* (AIC) [3] is used as the optimisation criterion. The equation for AIC is given in Eq. (6.7).

$$\mathrm{AIC} = 2(k) - 2log(L) \tag{6.7}$$

where $L$ is the maximum value of the likelihood function for the model, and $k$ is the number of estimated parameters in the model. Each dataset is trained with all twenty-five ARMA models and the AIC value for each model on each dataset is calculated. The ARMA model with the lowest average AIC value is selected. Table 6.6 shows the average AIC values for all twenty-five ARMA models with different orders of model parameters. As can be observed from Table 6.6, the lowest AIC value obtained is $(5.794) \times (10^5)$ when $p = 4$ and $q = 4$. Based on this result, the most suitable ARMA model to be used for one-step ahead prediction was found to be ARMA (4, 4).

3. ***Single Layer Linear Network Predictor (LN)*** [21, 45]
LN is derived from a Single Layer Perceptron architecture in an Artificial Neural Network. It is the simplest form of neural network as there is no hidden layer in the LN, and the connections between nodes do not form a cycle. In the LN method, the next data $\hat{Y}_{t+1}$ is predicted as a linear combination using data multiplied by a set of weight vectors represented by $v_0, v_1, \ldots, v_{W-1}$. The amount of previous data used in the one-step ahead prediction is determined by the size of the sliding window, denoted by $W$. This can be mathematically

**Table 6.6** Average Akaike information criterion (AIC) for different orders of ARMA model

| Model parameter | Average AIC | | | | |
|---|---|---|---|---|---|
| (p, q) | q = 0 | q = 1 | q = 2 | q = 3 | q = 4 |
| p = 0 | NIL | 6.541E+05 | 6.386E+05 | 6.233E+05 | 6.140E+05 |
| p = 1 | 6.059E+05 | 5.876E+05 | 5.820E+05 | 5.819E+05 | 5.806E+05 |
| p = 2 | 5.988E+05 | 5.824E+05 | 5.819E+05 | 5.817E+05 | 5.804E+05 |
| p = 3 | 5.875E+05 | 5.824E+05 | 5.815E+05 | 5.807E+05 | 5.803E+05 |
| p = 4 | 5.858E+05 | 5.800E+05 | 5.799E+05 | 5.799E+05 | **5.798E+05** |

expressed using Eq. (6.8):

$$\hat{Y}_{t+1} = \frac{\left(\sum_{i=0}^{W-1} v_i Y_{t-i}\right)}{\sum_{i=0}^{W-1} v_i} \tag{6.8}$$

This defines the relationship between the sliding window $Y_{t-W}, \ldots, Y_t$ and the predicted value of $\hat{Y}_{t+1}$. Following [45], we have assigned the weight vectors as $1, 2, \ldots, v$ with the weight vector inversely proportional to the distance between each point in the sliding window, that is, the further point $Y_t$ is from $\hat{Y}_{t+1}$, the smaller the weight vector.

Following the steps taken in determining the optimal values for parameters in SES and ARMA, finding the optimal window size, $W$, involves using all nine datasets. For each dataset, the value $W$ is varied between 1 and 10 in increments of 1. The MAE scores obtained for each $W$ size from all nine datasets are then averaged and the value $W$ that produces the lowest MAE score is selected. From the result in Table 6.7, the lowest average MAE obtained was 0.50 when $W = 3$.

A comparison is made between three different forecasting methods, namely SES, ARMA and LN. The optimal model for each method was obtained using Dataset 8 as training data. Besides using Dataset 1 as a validation dataset, the optimal models were also verified against Dataset 3, Dataset 5 and Dataset 9, which were used as validation datasets. Table 6.8 shows the performance of each dataset forecasting method against four different validation datasets.

The MAE obtained from Dataset 8 for all three forecasting methods are used as the baseline. For example, the baseline MAE for SES method is 0.526. When the forecast model is used to validate Dataset 1, the MAE obtained is 1.91, an increase of 3.63%. This indicates that the forecast model is not overfitting or underfitting. A model that is underfit will have high training and high validation error, while an overfit model will have extremely low training error but a high validation error. The results in Table 6.8 show a low validation error, less than 5% in all datasets. From the results shown in Table 6.8, all three forecasting methods are comparable with one another. However, the forecast model developed using the ARMA provides the lowest MAE score on all validation datasets compared to the SES and LN methods. This indicates that ARMA (4,4) creates a better forecast model for one-step ahead prediction.

#### 6.4.2.3 Measurement of Deviation and Anomaly Classification

Once the next data point has been predicted using either the SES, ARMA or LN methods, the next step is to define a measure of how much the observed behaviour of the time-series deviates from the expected pattern and if the observed value falls outside the defined threshold, it is classified as anomalous. For this purpose, a different Dijkstra dataset is used—one that has not been used for training and

**Table 6.7** Mean absolute error (MAE) c $W$ in a LN model

| Window size, (W) | MAE | | | | | | | | | Average MAE |
|---|---|---|---|---|---|---|---|---|---|---|
| | Dataset 1 | Dataset 2 | Dataset 3 | Dataset 4 | Dataset 5 | Dataset 6 | Dataset 7 | Dataset 8 | Dataset 9 | |
| 1 | **0.4197** | 0.4028 | 0.4028 | 0.5548 | 0.5628 | 0.5637 | 0.5732 | 0.5467 | 0.5472 | 0.51 |
| 2 | 0.4208 | **0.4003** | **0.4003** | 0.5527 | 0.5573 | 0.5621 | 0.5732 | 0.5425 | 0.5447 | 0.51 |
| 3 | 0.4264 | 0.4042 | 0.4042 | **0.5456** | **0.5491** | **0.5537** | **0.5667** | **0.5342** | **0.5386** | **0.50** |
| 4 | 0.4367 | 0.4134 | 0.4134 | 0.5494 | 0.5518 | 0.5554 | 0.5695 | 0.5365 | 0.5415 | 0.51 |
| 5 | 0.4491 | 0.4233 | 0.4233 | 0.5571 | 0.5581 | 0.5610 | 0.5747 | 0.5417 | 0.5476 | 0.52 |
| 6 | 0.4608 | 0.4322 | 0.4322 | 0.5631 | 0.5620 | 0.5647 | 0.5784 | 0.5456 | 0.5519 | 0.52 |
| 7 | 0.4710 | 0.44 | 0.44 | 0.5715 | 0.5685 | 0.5706 | 0.5846 | 0.5519 | 0.5587 | 0.53 |
| 8 | 0.4812 | 0.4473 | 0.4473 | 0.5806 | 0.5757 | 0.5771 | 0.5915 | 0.5584 | 0.5658 | 0.54 |
| 9 | 0.4909 | 0.4545 | 0.4545 | 0.5897 | 0.5830 | 0.5837 | 0.5982 | 0.5652 | 0.5729 | 0.54 |
| 10 | 0.500 | 0.46 | 0.46 | 0.5985 | 0.5901 | 0.5899 | 0.6047 | 0.5717 | 0.5797 | 0.55 |

**Table 6.8** Comparison between different forecasting methods in One-Step Ahead Prediction

| Training data (Dataset 8) | MAE (Dataset 1) | MAE (Dataset 3) | MAE (Dataset 5) | MAE (Dataset 9) |
|---|---|---|---|---|
| SES ($\alpha$= 0.6) MAE = 0.526 | 1.91(+3.63%) | 2.12(+4.03%) | 1.36(+2.59%) | 1.42(+2.70%) |
| ARMA(4,4) MAE = 0.592 | 1.86(+3.14%) | 2.06(+3.48%) | 1.32(+2.23%) | 1.38(+2.33%) |
| LN (W = 3) MAE = 0.534 | 1.87(+3.50%) | 2.08(+3.90%) | 1.34(+2.51%) | 1.39(+2.60%) |

which contains anomalies. Two different methods have been selected to measure the deviation between the observed data from the expected data, namely: (a) Residual Distribution; and (b) Prediction Interval, which will be explained further below.

1. ***Residual Distribution (RD)***: This method is adapted from Galvas [17] where the residual at specific time $t$ is used to define the deviation between predicted value and observed (or actual) value. The distributions of residuals from the SES, ARMA and LN methods follow a Cauchy distribution in which the mean is undefined and the variance is infinite. However, it is possible to calculate the residual average of a sample, denoted by $n$, in a Cauchy distribution [35], and from there, determine how many standard deviations away from the average of the forecast residual the current residual is lying. The equation for the Residual Distribution, (6.11), consists of two main components—the residual average, $\bar{e}$, and the residual variance, $\sigma^2$:

$$\text{Residual Average,} \quad \bar{e} = \frac{\sum_{i=1}^{n} e_i}{n} \tag{6.9}$$

$$\text{Residual Variance,} \quad \sigma^2 = \frac{\sum_{i=1}^{n} e_i^2}{n} - \bar{e}^2 \tag{6.10}$$

$$\text{Residual Distribution,} \quad z = \frac{e_t - \bar{e}}{\sqrt{\sigma^2}} \tag{6.11}$$

   The threshold rule, denoted as $z_{thresh}$, is defined as the distance of the forecast error from the residual average in terms of standard deviations to decide if the observed value should be marked as normal or anomalous. $z_{thresh}$ is varied between 1 and 10. After measuring the deviation between the predicted and observed values, if $z > z_{thresh}$, the observed value is marked as anomalous.
2. ***Prediction Interval (PI)*** [8]: The second method is a prediction interval which is commonly used in regression analysis. It is an estimate of the range in which predicted values will fall with a certain probability. The prediction interval describes the uncertainty for a single specific value where the uncertainty comes from the errors in the model itself and noise in the input data and provides the probabilistic upper and lower bounds based on the estimate of a predicted

variable. If the observed data falls between the upper and lower bounds, it is considered to be normal and if it falls outside those bounds, it will be marked as anomalous. To calculate the upper and lower bounds for a new predicted data point, $\hat{Y}_{t+1}$, Eq. (6.12) is used:

$$\begin{aligned} z_{upper} &= \hat{Y}_{t+1} + z_{bound} \\ z_{lower} &= \hat{Y}_{t+1} - z_{bound} \end{aligned} \tag{6.12}$$

where

$$z_{bound} = P_I * \sqrt{MSE}\sqrt{1 + \frac{1}{n} + \frac{\left(\hat{Y}_{t+1} - \bar{y}\right)^2}{\sum_{i=1}^{n}(e_i)^2}},$$

$$\text{MSE} = \sqrt{\frac{\sum_{i=1}^{n}(e_i)^2}{n}},$$

$$\bar{y} = \frac{1}{W} \cdot \sum_{i=t-W}^{t-1} Y_i$$

$n$ is the sample size up to time $t$, and $P_I$ represents the $100\%(1 - a, df)$ of the *Students T-distribution* with $df$ degrees of freedom [32, 39]. The critical value of $P_I$ reflects the confidence associated with the calculation of upper and lower bounds of $\hat{Y}_{t+1}$. The confidence level is varied between 80% and 97.5% and the value of $df$ between 1 and 3. The observed data point is considered normal if it satisfies the condition $z_{lower} < Y_t < z_{upper}$, else it is deemed anomalous.

#### 6.4.2.4 Evaluation of Early Detection and Prediction Algorithm

The early detection and prediction algorithm consists of three stages. The first stage is to predict the next data point using either SES, ARMA or LN one-step ahead prediction as discussed in Sect. 6.4.2.2. The second stage is to measure how much the observed data has deviated from the defined threshold, which has been discussed in Sect. 6.4.2.3. And finally, in the third stage, if the measurement of deviation does not satisfy the threshold rule, the observed data point is marked as anomalous.

In order to evaluate the effectiveness of the early detection and prediction algorithm, two evaluation metrics are used: accuracy and detection time. Calculation of both accuracy and detection time will be done using True Positives (TP), True Negatives (TN), False Positives (FP) and False Negatives (FN) as described in Table 6.9. TP refers to an anomalous observation, while TN refers to a normal observation. TP and TN are the ideal situations where data points are detected and identified correctly, while FP and FN are undesirable cases which need to be kept

**Table 6.9** Confusion matrix for early detection of anomalous behaviour

| | Detection | |
|---|---|---|
| Outcome | Anomalous | Non-Anomalous |
| Anomalous | True Positives (TP) (data points that are anomalous and identify as anomalous) | False Negatives (FN) (data points that are anomalous but identify as normal) |
| Non-Anomalous | False Positives (FP) (data points that are normal but identify as anomalous) | True Negatives (TN) (data points that are normal and identify as normal) |

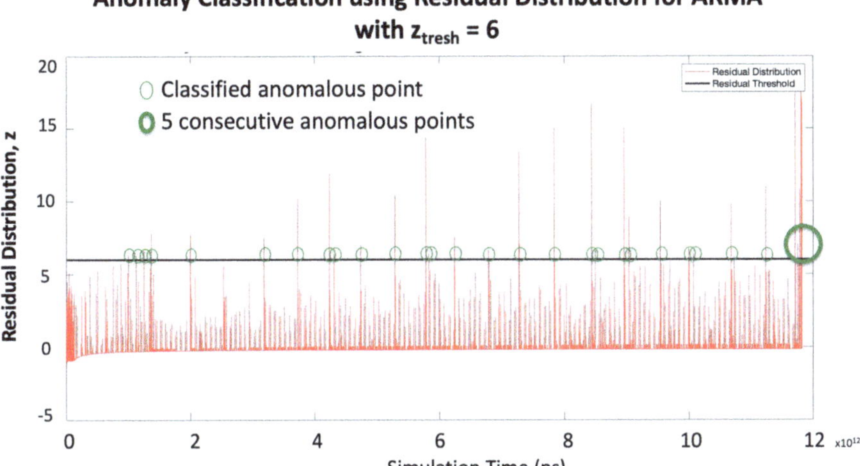


**Fig. 6.13** Anomaly classification using residual distribution

to a minimum. The formulae to calculate detection time and accuracy are given in Eqs. (6.13) and (6.14).

$$\text{Detection time} = (TP + FN) * \text{Logging Interval} \tag{6.13}$$

$$Accuracy = \frac{TP + TN}{(TP + FN + TN + FP)} \tag{6.14}$$

A good anomaly detection model should maximise the number of correct detections and keep false detections as low as possible [17], however, the main objective of the detection model is to be able to predict failure at the earliest time. Therefore, the detection time is a key attribute. As discussed earlier, the maximum detection interval has to be less than the maximum 4000 μs. Figures 6.13 and 6.14 show how anomalies are classified using either the Residual Distribution or Prediction Interval as the anomaly classification method after using the Autoregressive Moving Average for one-step-ahead prediction.

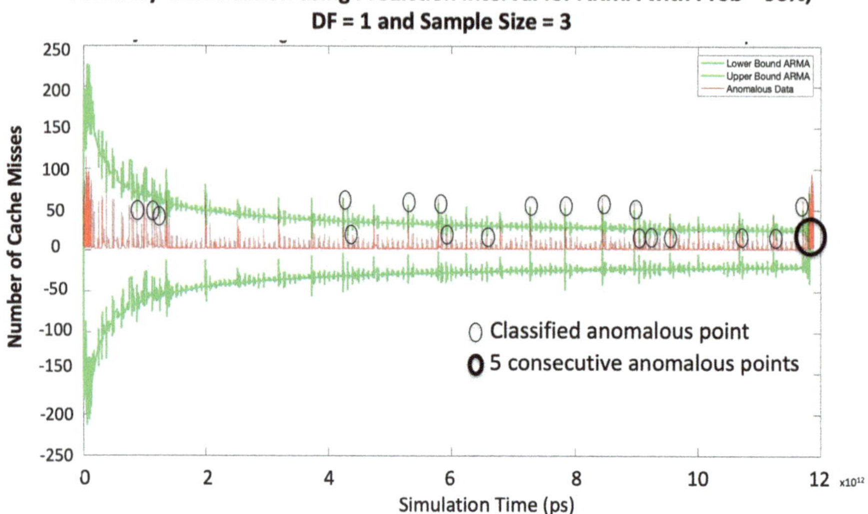


**Fig. 6.14** Anomaly classification using prediction interval

**Table 6.10** Analysis using SES, ARMA and LN method for number of successive anomalies, C = 5 using residual distribution

| Prediction methods | SES (a = 0.7) | | | ARMA (4,4) | | | LN (W=3) | | |
|---|---|---|---|---|---|---|---|---|---|
| Parameter ($z_{thresh}$) | 5 | 6 | 7 | 5 | 6 | 7 | 5 | 6 | 7 |
| TP | 17 | 15 | 21 | 25 | 19 | 23 | 20 | 18 | 21 |
| FN | 48 | 50 | 95 | 40 | 46 | 93 | 45 | 47 | 95 |
| FP | 216 | 132 | 68 | 163 | 89 | 47 | 205 | 106 | 60 |
| TN | 117,940 | 118,024 | 118,088 | 117,993 | 118,067 | 118,109 | 117,951 | 118,050 | 118,096 |
| Accuracy | 99.78% | 99.85% | 99.86% | 99.83% | 99.89% | 99.88% | 99.79% | 99.87% | 99.87% |
| Detection time $\mu$s | 325 | 325 | 580 | 325 | 325 | 580 | 325 | 320 | 580 |

In Fig. 6.13 the residual threshold (or $z_{thresh}$) is set at 6, and values that exceed the threshold rule are marked as anomalous. In Fig. 6.14, the upper and lower bounds ($z_{upper}$ and $z_{lower}$) become the boundary threshold and any points beyond this threshold are marked as anomalous. The detection and prediction continue until five consecutive anomalies are detected. As can be observed from these two figures, the majority of the points lie well below the thresholds. In Fig. 6.14, the upper and lower bounds generated using the Prediction Interval provide a good envelope for the actual data, but at least 20 $\mu$s is required for the calculation to stabilise. This means that if anomalies happen in the start-up phase of the execution of a program, this method would not be able to detect those anomalies and would not then predict a failure.

Table 6.10 shows the top three results for each method using Residual Distribution. The sampling interval for the data is set at 5 μs and the best performance is achieved with $z_{thresh} = 6$, and the earliest detection time is 325 μs (or ≈ 82000 clock cycles) after a fault has been manifested as an error, a result that is significantly better than our previous work [44]. The number of anomalous points, TP, that are correctly detected is between 15 and 19, while the non-anomalous points, TN, that are correctly identified lie between 118,050 and 118,067, giving an accuracy between 99.85% and 99.89%. The number of missed anomalies and false alarms (FN and FP) is between 135 and 182.

In Table 6.11, the top three results for each prediction method using Prediction Interval are presented. The sampling interval is 5 μs and the best results were achieved with $PI = 3.078$ and with the lowest detection time of 325 μs. Although the results using $PI = 2.920$ and $PI = 2.353$ have the same detection time of 325 μs, the FP number is much higher compared to the results achieved with $PI = 3.078$. The number of anomalous points, TP, that are correctly detected is between 15 and 19, while the non-anomalous points, TN, that are correctly identified lie between 118,058 and 118,072, giving an accuracy between 99.88% and 99.89%.

These two results show that ARMA provides slightly higher accuracy than SES and LN. Although the value of TN is a little higher, the ARMA method has reduced FN and FP. The difference between Residual Distribution and Prediction Interval methods can be seen from number of normal data points wrongly identified as anomalous (FP). The optimised parameters obtained from this analysis will be used in the experimental validation of the detector.

### 6.4.3 Experimental Validation of the Detector

In previous research that used HPCs to detect anomalous behaviour, the data collected was sent for offline analysis and anomalies were only detected after a failure had happened [1, 9]. In contrast to that, this detector can predict failures in real-time. The key difference lies in the ability of the detector to detect and predict within a certain number of clock cycles and hence, to prevent the system from entering into a failure state. In this section, the implementation of the detector is presented.

The idea of using a dedicated hardware processor to detect anomalous behaviour in the main processor is aimed at achieving a quick response for detection and prediction with minimal performance overhead. Rather than placing the detector on the main core in a multicore microcontroller, it is designed to be placed on secondary core to ensure no overhead is imposed on the main core running the application. Following the design guidelines as proposed in [11], the main core and secondary core have been designed to have private caches and private memories. This is to ensure that the HPC data from the main core that uses the number of cache misses as its PME will not be compromised due to the presence of a secondary core.

**Table 6.11** Analysis using SES, ARMA and LN method for number of successive anomalies, C = 5 using prediction interval

| Methods | SES ($\alpha = 0.7$) | | | ARMA (4,4) | | | LN (W=3) | | |
|---|---|---|---|---|---|---|---|---|---|
| Parameter PI (prob., df) | PI = 3.078 (prob = 90%, df = 1, W = 3) | PI = 2.920 (prob = 95%, df = 2, W = 3) | PI = 2.353 (prob = 95%, df = 3, W = 5) | PI = 3.078 (prob = 90%, df = 1, W = 3) | PI = 2.920 (prob = 95%, df = 2, W = 3) | PI = 2.353 (prob = 95%, df = 3, W = 5) | PI = 3.078 (prob = 90%, df = 1, W = 3) | PI = 2.920 (prob = 95%, df = 2, W = 3) | PI = 2.353 (prob = 95%, df = 3, W = 5) |
| TP | 15 | 15 | 15 | 19 | 22 | 22 | 17 | 18 | 18 |
| FN | 50 | 50 | 50 | 46 | 43 | 43 | 48 | 47 | 47 |
| FP | 88 | 100 | 102 | 84 | 108 | 114 | 98 | 127 | 135 |
| TN | 118,068 | 118,056 | 118,054 | 118,072 | 118,048 | 118,042 | 118,058 | 118,029 | 118,021 |
| Accuracy | 99.88% | 99.87 % | 99.87 % | 99.89 % | 99.87 % | 99.87 % | 99.88 % | 99.85% | 99.85 % |
| Detection time (μs) | 325 | 325 | 325 | 325 | 325 | 325 | 325 | 325 | 325 |

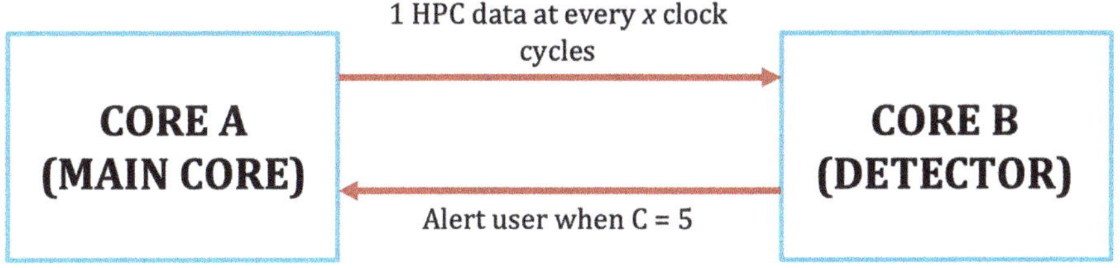


**Fig. 6.15** Hardware-based detector utilising multicore architecture

#### 6.4.3.1 Design of the Detector

Figure 6.15 illustrates how the detector is designed in such a way that the secondary processor can receive the PME counts from the main processor with minimal overhead to the main processor via the communication pipeline. The important consideration in this design is the inter-core communication pipeline where the HPC data from Core A can be sent via a dedicated pipeline to Core B as shown in Fig. 6.15. It is important to have a dedicated pipeline to ensure that the HPC data from Core A will not be compromised, which can affect the detection and prediction of potential failure.

Figure 6.16 shows the overall execution flow between the main core, Core A, and the secondary core, Core B. Core A which functions as the main core starts up the whole microcontroller, initialises the memory, peripherals and stack pointers. Core A will then turn on Core B and initialises the inter-core communication pipeline to Core B. Core B then loads the detector program. Core A will run the application and send one HPC data point every 5 μs to Core B. Core B, prior to receiving the HPC data, will perform one-step ahead prediction and predict of the next data. Once the actual data (or observed data) is available, Core B will measure the deviation between predicted data and actual data. If it exceeds the threshold, the actual data will be marked as anomalous. If there are five anomalous data points detected consecutively, Core B sends an interrupt to Core A. Core A, upon receiving the interrupt, will close the inter-core communication pipeline with Core B, halt the application and raise an alarm for potential failure.

The validation is performed on a workstation running Ubuntu 16.04 LTS with Intel Core i5-5257U CPU operating at 2.70 GHz and 11.1 GB of memory. In the experimental setup, Core A and Core B are simulated using two stand-alone programs. A stream of HPC data from a benchmark is used as an input to Core A. The program for Core B consists of the early detection and prediction program to predict and classify the stream of HPC data coming from Core A. The implementation of inter-core communication pipeline between Core A and Core B is realised using a *named pipe* (FIFO). The time granularity has been fixed such that a single item of HPC data is transferred from Core A to Core B every 5 μs.

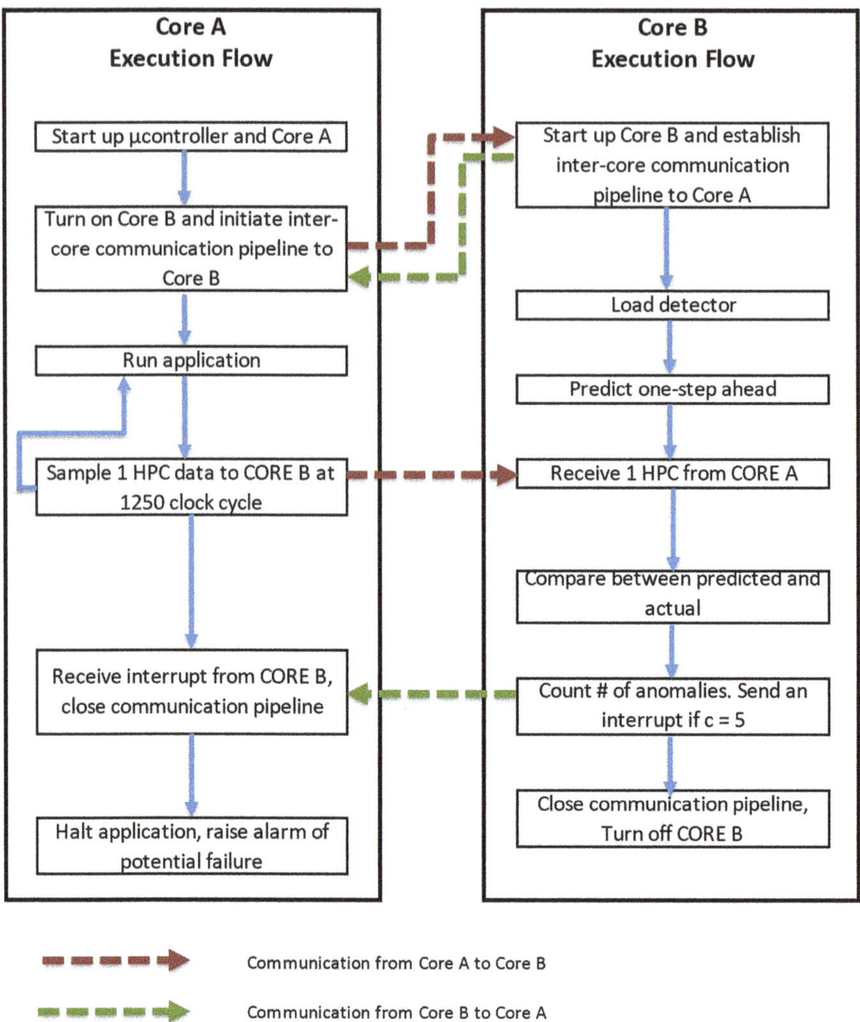


**Fig. 6.16** Overall execution flow between main core, Core A, and secondary core, Core B

For each HPC data received from Core A, Core B determines if the current point is deemed anomalous. Core B keeps the count of anomalous points it has detected and if Core B detects five anomalous points consecutively, it sends an interrupt to Core A via another communication pipeline to notify Core A of the potential failure. This is to ensure that the pipeline used to send HPC data from Core A to Core B does not need to wait and check for any alarm from Core B, which could impede the detection process. Upon receiving the interrupt, the program in Core A closes the FIFO and displays an error message notifying the user of a potential failure.

The optimal parameters determined in Sect. 6.4.2 are used in the early detection and prediction algorithms.

Three different benchmarks have been chosen to validate the detector. The benchmarks chosen are Dijkstra, FFT and Bitcount benchmarks, each from a different suite. These benchmark applications have been injected with a single bit-flip randomly at various stages of the pipeline. As Core A is running a benchmark application, it sends one piece of HPC data at every 1250 clock cycles through the communication pipeline to Core B. Six different detectors were implemented with each detector running a combination of one-step ahead prediction with anomaly classification techniques. In Core B, the detector will classify if the current point is anomalous and raise the alarm of the impending failure of Core A if five anomalous points are detected consecutively. The interrupt is transmitted to Core A via another dedicated communication pipeline. This is to ensure that the pipeline that is being used to send an HPC data from Core A to Core B does not need to wait and check for any alarm from Core B, which could impede the detection process.

#### 6.4.3.2 Simulation Results

The six different detectors were tested on three benchmarks, namely Dijkstra, FFT and Bitcount benchmarks. These benchmarks are tasked to run at the same clock speed and provide HPC data at the same sampling interval of 5 μs. It should be emphasised that the early detection and prediction algorithm was developed using fault-free datasets from Dijkstra benchmark. For the validation of the detector, two additional benchmarks were used, the Bitcount and FFT benchmarks.

In Dijkstra benchmark, Table 6.12 shows how all six detectors have managed to predict a potential failure in the system at 2595 μs, 350 μs, 2595 μs, 2425 μs, 405 μs and 1075 μs, respectively. This means, the detector core had successfully predicted a potential failure of the main core before actual failure occurs. As for Bitcount benchmark, Table 6.13 shows the result of the detector detecting anomalies and predicting a potential failure with all detectors managed to detect the anomalies and predict potential failure with the detection time ranging from 525 μs to 2450 μs, while for FFT benchmark, Table 6.14 shows the results for each detector in detecting

**Table 6.12** Detection time for Dijkstra benchmark anomalous dataset

| Anomaly classification | Residual distribution | | | Prediction interval | | |
|---|---|---|---|---|---|---|
| One-step ahead prediction | SES | ARMA | LN | SES | ARMA | LN |
| Fault injection at time (s): | 6.6587 | | | 6.6587 | | |
| System crash at time (s): | 6.6636 | | | 6.6636 | | |
| Start of anomalous behaviour at time (s): | 6.6597 | | | 6.6597 | | |
| Anomalies detected at time (s): | 6.6623 | 6.6600 | 6.6620 | 6.6622 | 6.6601 | 6.6608 |
| Detection time at (us): | 2595 | 350 | 2595 | 2425 | 405 | 1075 |

**Table 6.13** Detection time for Bitcount benchmark anomalous dataset

| Anomaly classification | Residual distribution | | | Prediction interval | | |
|---|---|---|---|---|---|---|
| One-step ahead prediction | SES | ARMA | LN | SES | ARMA | LN |
| Fault injection at time (s): | 6.6756 | | | 6.6756 | | |
| System crash at time (s): | 6.6806 | | | 6.6806 | | |
| Start of anomalous behaviour at time (s): | 6.6766 | | | 6.6766 | | |
| Anomalies detected at time (s): | 6.6791 | 6.6771 | 6.6790 | 6.6791 | 6.6772 | 6.6790 |
| Detection time at (us): | 2465 | 525 | 2420 | 2450 | 545 | 2375 |

**Table 6.14** Detection time for FFT benchmark anomalous dataset

| Anomaly classification | Residual distribution | | | Prediction interval | | |
|---|---|---|---|---|---|---|
| One-step ahead prediction | SES | ARMA | LN | SES | ARMA | LN |
| Fault injection at time (s): | 7.1101 | | | 7.1101 | | |
| System crash at time (s): | 7.1161 | | | 7.1161 | | |
| Start of anomalous behaviour at time (s): | 7.1121 | | | 7.1121 | | |
| Anomalies detected at time (s): | 7.1157 | 7.1130 | 7.1141 | 7.1144 | 7.1137 | 7.1137 |
| Detection time at (us): | 4025 | 930 | 1960 | 2270 | 1610 | 1635 |

the anomalous behaviour and predicting a potential failure, with the detection time ranges from 930 μs to 4025 μs.

As shown in Fig. 6.17, the detector is able to detect and predict potential failure before the system crashes for all three benchmarks. "Time to Failure" indicates the time from when the system started to behave anomalously until the failure occurred, while "Time to Detect" is the time the detector took to predict a potential failure. From Fig. 6.17, the fastest time to detect was obtained using ARMA. A quicker detection time means more time to undertake any preventive or corrective measures. For measuring the deviation between the predicted point and the actual point, both Residual Distribution and Prediction Interval are comparable. The only downside to using Prediction Interval for anomaly classification is that it requires at least 20 μs for the calculation to stabilise, during which time a fault is not detectable. These results show that the early detection and prediction algorithm developed is generalisable and can be applied to other benchmarks.

### 6.4.4 *Performance Analysis*

Next, the performance of the detector is evaluated by calculating the total time to predict one data point, compare and measure the deviation between the predicted and observed data, and decide whether it is anomalous or not. This total execution time, $T$, is calculated using the formula in (6.15) where $I$ is the total number of instructions and $CPI$ is the Cycles per Instruction.

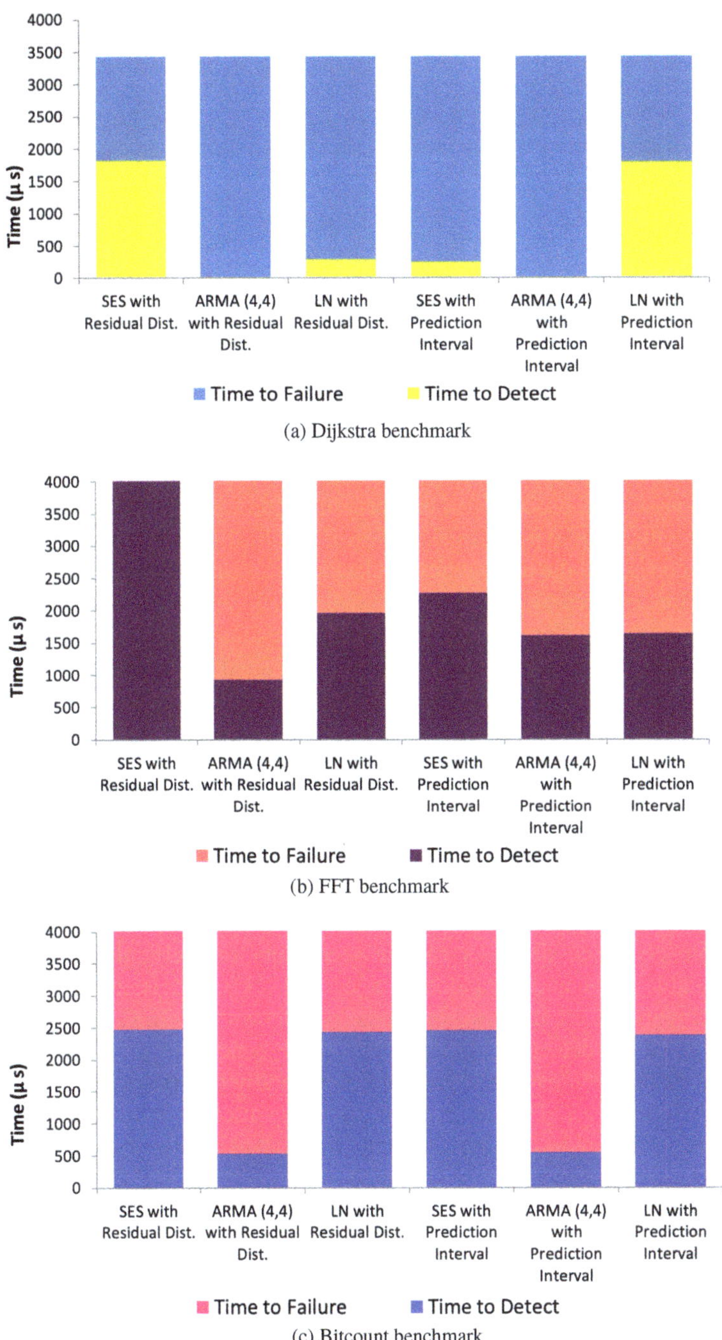


(a) Dijkstra benchmark



(b) FFT benchmark



(c) Bitcount benchmark

**Fig. 6.17** Prediction performance for three different benchmarks. (**a**) Dijkstra benchmark. (**b**) FFT benchmark. (**c**) Bitcount benchmark

**Table 6.15** Performance in execution time of each method measured on an Intel architecture

| Method | Total instructions (I) | Cycles per instruction (CPI) | Total execution time (T)-ns |
|---|---|---|---|
| SES + RD | 250 | 1.8780 | 1878 |
| ARMA(4,4) + RD | 311 | 1.7186 | 2138 |
| LN + RD | 243 | 1.8992 | 1846 |
| SES + PI | 334 | 1.7006 | 2272 |
| ARMA(4,4) + PI | 384 | 1.6810 | 2582 |
| LN + PI | 330 | 1.8273 | 2412 |

$$T = I * CPI * \text{CPU Clock Cycle} \tag{6.15}$$

The detector is placed on Core B rather than Core A to ensure no overhead is imposed on the main core running the benchmarks. As the main benchmarks were executed on an Intel processor in the form of single-tasking system, we placed the detector on the same type of Intel processor. Each of the programs is broken into assembly codes, and the throughput and latencies for each instruction are calculated based on Intel processor data [15]. The CPU clock period is 1/Clock Rate where the clock rate is 250 MHz. Table 6.15 shows the results for each method.

As Table 6.15 shows, the total execution time for detection ranges from around 1800–2600 ns, well below the sampling period of 5000 ns (sampling every 1250 clock cycles at a clock rate of 250 MHz). In other words, as a data point is sampled, the detector is able to determine if it is normal or anomalous. The total number of instructions using Prediction Interval is higher than for Residual Distribution as the computation for lower and upper bounds is more complex. The CPI for methods using Residual Distribution is slightly higher as there is more dependency in the detection and prediction algorithm, hence there are more latencies.

The size of the detector in software terms is measured by the size of the executable code and data. As can be seen from Fig. 6.18, the combined size of executable code and data does not exceed 2 kB and is almost as light as the benchmarks used in this experiment, as shown in Fig. 6.19. The size and complexity of the detector is independent of the benchmarks. As ARMA has the most instructions compared to SES and LN, naturally the size of the detector using ARMA will be bigger. The size of the detector that uses Prediction Interval method is also bigger by almost 500 bytes compared to that using Residual Distribution.

### 6.4.5 Summary

From these analyses, it is observed that all the techniques detected anomalous behaviour well before the system failed, but that ARMA+RD is the fastest, Fig. 6.17. The RD methods are faster than PI for evaluating a single data point, Table 6.15 and

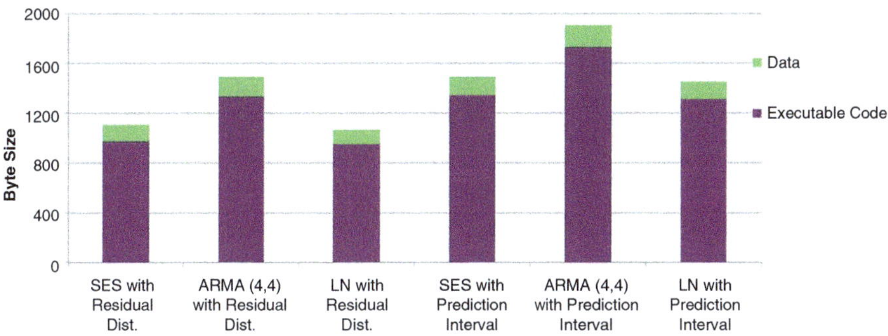


**Fig. 6.18** Size of EDAB detector in bytes

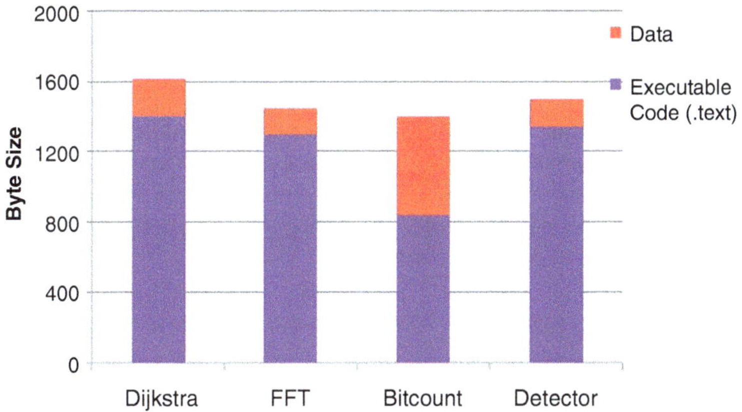


**Fig. 6.19** Size of detector in comparison with other benchmarks

have slightly smaller code sizes, Fig. 6.18. Thus, by a small margin, ARMA+RD is the best choice for prediction and detection of anomalies.

Placing the detector on a secondary core means there will be no additional hardware required on the main core. The main core utilises the existing hardware performance counter in its own core and send the data to the secondary core for detection of anomalous behaviour. Based on performance analysis, the detector can be deemed as a lightweight detector since the size is below 2 kB. Although the experiments are done in the form of simulations, other work has demonstrated that using a secondary processor core to monitor the main processor core's HPCs is possible [1]. However, compared to our work which uses real-time streaming HPC data to detect for anomalies, they captured the total count of HPC data after the application has completed and performed the analysis offline to determine if the application is benign or anomalous.

## 6.5 Conclusion

This chapter presents the novel algorithm to predict potential failure in real-time by monitoring and detecting anomalous behaviour in a processor. Embedded systems have limitations and constraints concerning hardware resources, speed, power and memory size. Therefore, the algorithm for error detection and prediction of potential failure must be lightweight with minimal computational complexity and do not require any preprocessing on the data. Statistical methods are preferred over machine learning algorithms as these methods not only satisfy the above criteria, but statistical methods have also been found to outperformed machine learning algorithms in terms of forecasting accuracy.

Based on the findings of the case study presented in this chapter, it can be concluded that it is possible to predict a potential failure in the embedded system by monitoring the system for any anomalous behaviour. The lightweight detector proposed is suitable to be used in multicore microcontrollers and there is no additional cost imposed on the main core running the application. This novel algorithm for early detection and prediction of potential failure in a processor has proven to work even on benchmarks that were not used for training and testing. This algorithm can complement existing fault forecasting and fault tolerance techniques and will contribute to a better protection strategy for microprocessors, especially those that are used in embedded computing.

**Acknowledgments** This work has been partly supported by Microsoft Azure Research Award number CRM: 0518905.

## References

1. M.F.B. Abbas, S.P. Kadiyala, A. Prakash, T. Srikanthan, Y.L. Aung, Hardware performance counters based runtime anomaly detection using SVM, in TRON Symposium (TRONSHOW) (2017) , pp. 1–9. https://doi.org/10.23919/TRONSHOW.2017.8275073
2. S. Ahmad, A. Lavin, S. Purdy, Z. Agha, Unsupervised real-time anomaly detection for streaming data. Neurocomputing **262**(Supplement C), 134–147 (2017). https://doi.org/10.1016/j.neucom.2017.04.070. Online Real-Time Learning Strategies for Data Streams
3. H. Akaike, A new look at the statistical model identification. IEEE Trans. Autom. Control **19**(6), 716–723 (1974). https://doi.org/10.1109/TAC.1974.1100705
4. P.C. Anderson, F.J. Rich, S. Borisov, Mapping the South Atlantic Anomaly continuously over 27 years. J. Atmos. Sol. Terr. Phys. **177**, 237–246 (2018). https://doi.org/10.1016/j.jastp.2018.03.015. Dynamics of the Sun-Earth System: Recent Observations and Predictions
5. A. Avizienis, Fundamental concepts of dependability. Comput. Oper. Res., 1–20 (2012)
6. M.B. Bahador, M. Abadi, A. Tajoddin, HPCMalHunter: behavioral malware detection using hardware performance counters and singular value decomposition, in Proceedings of the 4th International Conference on Computer and Knowledge Engineering (ICCKE) (2014), pp. 703–708. https://doi.org/10.1109/ICCKE.2014.6993402
7. V. Chandola, A. Banerjee, V. Kumar, Anomaly detection: A survey. ACM Comput. Surv. **41**(3), 15:1–15:58 (2009). https://doi.org/10.1145/1541880.1541882

8. C. Chatfield, *Prediction Intervals for Time-Series Forecasting* (Springer, Boston, 2001), pp. 475–494. https://doi.org/10.1007/978-0-306-47630-3_21
9. E. Chavis, H. Davis, Y. Hou, M. Hicks, S.F. Yitbarek, T. Austin, V. Bertacco, SNIFFER: a high-accuracy malware detector for enterprise-based systems, in Proceedings of the IEEE 2nd International Verification and Security Workshop (IVSW) (2017), pp. 70–75. https://doi.org/10.1109/IVSW.2017.8031547
10. M. Chiappetta, E. Savas, C. Yilmaz, Real time detection of cache-based side-channel attacks using hardware performance counters. Appl. Soft Comput. **49**(C), 1162–1174 (2016). https://doi.org/10.1016/j.asoc.2016.09.014
11. C. Cullmann, C. Ferdinand, G. Gebhard, D. Grund, C.M. (Burguiére), J. Reineke, B. Triquet, R. Wilhelm, Predictability considerations in the design of multi-core embedded systems, in Embedded Real Time Software and Systems Conference (2010), pp. 36–42
12. A. DeHon, N. Carter, H. Quinn, Final report of CCC cross-layer reliability visioning study, in *Full Report of Computing Community Consortium (CCC) Visioning Study*. Computing Community Consortium (CCC) Visioning Study, United States (2011). http://www.relxlayer.org/FinalReport?action=AttachFile&do=view&target=final_report.pdf
13. J. Dromard, G. Roudiére, P. Owezarski, Online and scalable unsupervised network anomaly detection method. IEEE Trans. Netw. Serv. Manag. **14**(1), 34–47 (2017). https://doi.org/10.1109/TNSM.2016.2627340
14. N.H. Duong, H.D. Hai, A semi-supervised model for network traffic anomaly detection, in Proceedings of the 17th International Conference on Advanced Communication Technology (ICACT) (2015), pp. 70–75. https://doi.org/10.1109/ICACT.2015.7224759
15. A. Fog, *4. Instruction Tables*. Software Optimization Resources. (Technical University of Denmark, Denmark, 2018). https://www.agner.org/optimize/instruction_tables.pdf
16. W.A. Fuller, *Introduction to Statistical Time Series* (Wiley, New York, 1976)
17. G. Galvas, *Time Series Forecasting used for Real-time Anomaly Detection on Websites*. Master's thesis (Faculty of Science, Vrije Universiteit, Vrije, 2016)
18. M. Goldstein, S. Uchida, A comparative evaluation of unsupervised anomaly detection algorithms for multivariate data. PLOS One, 1–31 (2016). https://doi.org/10.1371/journal.pone.0152173
19. M.R. Guthaus, J.S. Ringenberg, D. Ernst, T.M. Austin, T. Mudge, R.B. Brown, MiBench: a free, commercially representative embedded benchmark suite, in *Proceedings of the IEEE International Workshop Workload Characterization (WWC-4), WWC '01* (IEEE Computer Society, Washington, 2001), pp. 3–14. https://doi.org/10.1109/WWC.2001.15
20. R.S. Hammer, D.T. McBride, V.B. Mendiratta, Comparing reliability and security: concepts, requirements and techniques. Bell Labs Tech. J. **12**(3), 65–78 (2007). https://doi.org/10.1002/BLTJ.20250
21. D.J. Hill, B.S. Minsker, Anomaly detection in streaming environmental sensor data: a data-driven modeling approach. Environ. Model. Softw. **25**(9), 1014–1022 (2010). https://doi.org/10.1016/j.envsoft.2009.08.010
22. R.J. Hyndman, G. Athanasopoulos, *Forecasting: Principles and Practice*, 2nd edn. (OTexts, Melbourne, 2019). https://otexts.com/fpp2
23. M.S. Islam, W. Khreich, A. Hamou-Lhadj, Anomaly detection techniques based on kappa-pruned ensembles. IEEE Trans. Reliab. **67**(1), 212–229 (2018). https://doi.org/10.1109/TR.2017.2787138
24. R. Iyer, Z. Kalbarczyk, W. Gu, *Benchmarking the Operating System against Faults Impacting Operating System Functions* (Wiley, New York, 2008), pp. 311–339. https://doi.org/10.1002/9780470370506.ch15. http://dx.doi.org/10.1002/9780470370506.ch15
25. Y. Kawachi, Y. Koizumi, N. Harada, Complementary set variational autoencoder for supervised anomaly detection, in *IEEE International Conference on Acoustics, Speech and Signal Processing (ICASSP)* (2018), pp. 2366–2370. https://doi.org/10.1109/ICASSP.2018.8462181
26. A. Kumar, A. Srivastava, N. Bansal, A. Goel, Real time data anomaly detection in operating engines by statistical smoothing technique, in *Proceedings of the 25th IEEE Canadian Conference on Electrical and Computer Engineering (CCECE)* (2012), pp. 1–5. https://doi.

org/10.1109/CCECE.2012.6334876
27. E.W.L. Leng, M. Zwolinski, B. Halak, Hardware performance counters for system reliability monitoring, in *Proceedings of the IEEE 2nd International Verification and Security Workshop (IVSW)*, pp. 76–81 (2017). https://doi.org/10.1109/IVSW.2017.8031548
28. N.G. Leveson, C.S. Turner, An investigation of the Therac-25 accidents. Computer **26**(7), 18–41 (1993). https://doi.org/10.1109/MC.1993.274940
29. J.L. Lions, ARIANE 5—-flight 501 failure, in *Failure report, Independent Inquiry Board* (1996). https://esamultimedia.esa.int/docs/esa-x-1819eng.pdf
30. S. Makridakis, E. Spiliotis, V. Assimakopoulos, Statistical and machine learning forecasting methods: concerns and ways forward. PLoS ONE **13**(3), e0194889. PLoS ONE (2018). https://journals.plos.org/plosone/article/file?id=10.1371/journal.pone.0194889&type=printable
31. D.C. Montgomery, C.L. Jennings, M. Kulahci, *Introduction to Time Series Analysis and Forecasting*. Wiley Series in Probability and Statistics. (Wiley, New York, 2011). https://books.google.co.uk/books?id=-qaFi0oOPAYC
32. *NIST: Nist/sematech e-handbook of Statistical Methods* (2013). http://www.itl.nist.gov/div898/handbook/
33. K. Parasyris, G. Tziantzoulis, C.D. Antonopoulos, N. Bellas, GemFI: a fault injection tool for studying the behavior of applications on unreliable substrates, in *Proceedings of the 44th Annual IEEE/IFIP International Conference on Dependable Systems and Networks* (2014), pp. 622–629. https://doi.org/10.1109/DSN.2014.96
34. F. Petropoulos, S. Makridakis, V. Assimakopoulos, K. Nikolopoulos, 'Horses for Courses' in demand forecasting. Eur. J. Oper. Res. **237**, 152–163 (2014). https://doi.org/10.1016/j.ejor.2014.02.036
35. N.S. Pillai, X.L. Meng, An unexpected encounter with Cauchy and Lèvy. Ann. Statist. **44**(5), 2089–2097 (2016). https://doi.org/10.1214/15-AOS1407
36. Y. Sasaka, T. Ogawa, M. Haseyama, A novel framework for estimating viewer interest by unsupervised multimodal anomaly detection. IEEE Access **6**, 8340–8350 (2018). https://doi.org/10.1109/ACCESS.2018.2804925
37. H. Song, Z. Jiang, A. Men, B. Yang, A hybrid semi-supervised anomaly detection model for high-dimensional data. Comput. Intell. Neurosci. **2017**, 1–9 (2017). https://doi.org/10.1155/2017/8501683
38. L. Song, H. Liang, T. Zheng, Real-time anomaly detection method for space imager streaming data based on HTM algorithm, in *Proceedings of the IEEE 19th International Symposium on High Assurance Systems Engineering (HASE)* (2019), pp. 33–38. https://doi.org/10.1109/HASE.2019.00015
39. t-distribution table. http://www.sjsu.edu/faculty/gerstman/StatPrimer/t-table.pdf
40. S. Teoh, *RM142m RazakSAT Faulty After Just One Year, Says Federal Auditor* (The Malaysian Insider, Malaysian, 2011)
41. M. Toledano, I. Cohen, Y. Ben-Simhon, I. Tadeski, Real-time anomaly detection system for time series at scale, in *Proceedings of Machine Learning Research*. Workshop on Anomaly Detection in Finance (PMLR), vol. 71 (2018), pp. 56–65
42. V. Vercruyssen, W. Meert, G. Verbruggen, K. Maes, R. Bäumer, J. Davis, Semi-supervised anomaly detection with an application to water analytics, in *Proceedings of the IEEE International Conference on Data Mining (ICDM)* (2018), pp. 527–536. https://doi.org/10.1109/ICDM.2018.00068
43. N. Wehn, Reliability: a cross-disciplinary and cross-layer approach, in *Asian Test Symposium* (2011), pp. 496–497

44. L.L. Woo, M. Zwolinski, B. Halak, Early detection of system-level anomalous behaviour using hardware performance counters, in *Design, Automation Test in Europe Conference Exhibition (DATE)* (2018), pp. 485–490. https://doi.org/10.23919/DATE.2018.8342057
45. Y. Yu, Y. Zhu, S. Li, D. Wan, Time series outlier detection based on sliding window prediction. Math. Probl. Eng. **2014**(879736), Article ID 879736 (2014). https://doi.org/10.1155/2014/879736
46. J. Zhao, L. Xu, L. Liu, Equipment fault forecasting based on ARMA model, in *Proceedings of the International Conference on Mechatronics and Automation* (2007), pp. 3514–3518. https://doi.org/10.1109/ICMA.2007.4304129

# Index





B. Halak (ed.), *Hardware Supply Chain Security*,
https://doi.org/10.1007/978-3-030-62707-2

## H



## I



## K



## M

**T**



**V**



**W**

GPSR Compliance
The European Union's (EU) General Product Safety Regulation (GPSR) is a set of rules that requires consumer products to be safe and our obligations to ensure this.

If you have any concerns about our products, you can contact us on

ProductSafety@springernature.com

In case Publisher is established outside the EU, the EU authorized representative is:

Springer Nature Customer Service Center GmbH
Europaplatz 3
69115 Heidelberg, Germany

www.ingramcontent.com/pod-product-compliance
Ingram Content Group UK Ltd.
Pitfield, Milton Keynes, MK11 3LW, UK
UKHW021834270726
14058UKWH00001B/140
* 9 7 8 3 0 3 0 6 2 7 0 9 6 *